D0607018

LOST
DISCOVERIES

The Forgotten Science
of the Ancient World

LOST
DISCOVERIES

Colin Ronan

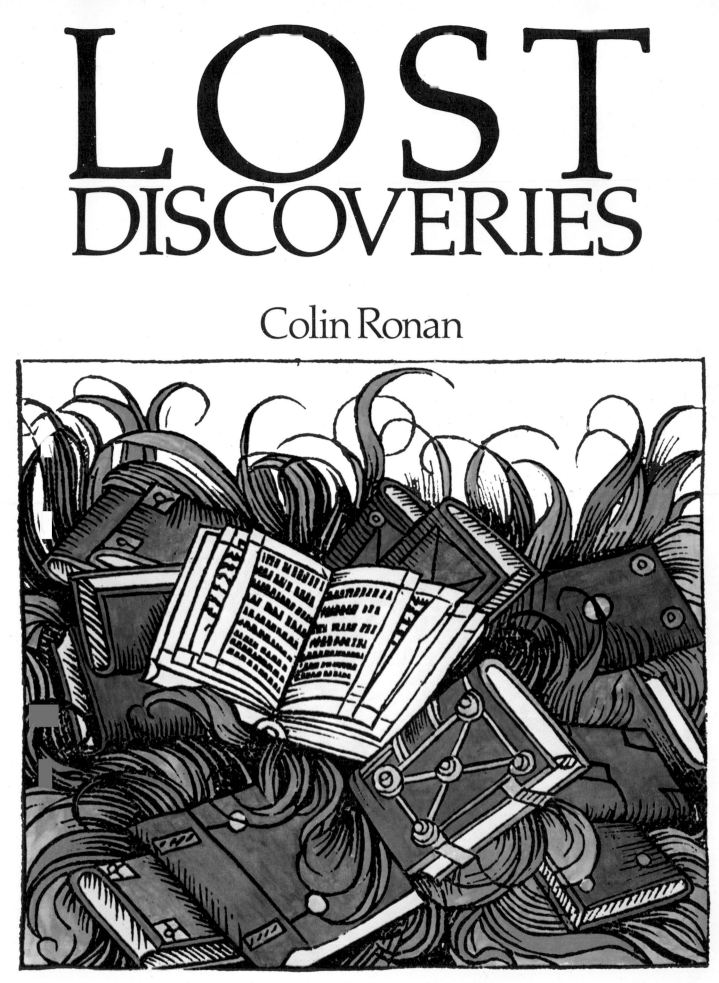

McGraw-Hill Book Company

New York·St.Louis·San Francisco·Toronto

Library of Congress Cataloging in Publication Data
Ronan, Colin A.
  Lost Discoveries
  1. Science, Ancient. 2. Inventions. 1. Title
  Q125.R742   509'.3   73–4968
ISBN 0–07–053597–3

Made by Roxby Press Productions
55 Conduit Street   London W1R ONY
Editor: Michael Wright
Picture Research: Ronan Picture Library/Anne Horton
Design & art direction: Dodd and Dodd
Printed in Great Britain by Morrison & Gibb Limited
London and Edinburgh

Introduction

CURIOSITY is as old as man himself. Stretching back into the dim recesses of the past, long before the first crude fumblings of recorded time, it is the impelling urge behind man's attempts to explain the natural world and find ways to control it. It was curiosity that, more than 300,000 years ago, led him to conquer fire and use it for warmth, for cooking, for smelting – and even for war. But curiosity alone was not enough. Without a means of recording and communicating discoveries and ideas, the accumulation of knowledge was limited. It remained so until the advent of writing.

Although the art of writing was invented 6,000 years ago, and used for accounting and receipts, it was a millennium and more before an alphabet evolved and the written word at last became the commerce of scholars. Records in stone, on clay tablets or on sheets of papyrus were then gathered to form the nucleus of libraries, and it is on these that our knowledge of past cultures to a great extent depends. But there was no guarantee of transmission; the written record was not proof against disasters. Earthquakes, hurricanes, flood and fire might ravage stone, clay and papyrus alike, and what was left could perish in war or be destroyed wilfully from fear or bigotry.

Yet, in spite of all these hazards, an amazing amount has come down to us, either in the form of the actual records or in copies and commentaries prepared from originals now lost. Archeologists have unearthed tombs and excavated cities; artists, historians, philologists, philosophers, scientists and technologists have applied their special skills to what has been found; and now we can trace much of the development of our civilization from its most important source in the eastern Mediterranean.

To do this is to construct a jig-saw from a vast collection of pieces, and even though not every fragment is in place, it has become clear that the civilizations of Crete, of Mesopotamia and of Egypt, all of which began to develop not quite 5,000 years ago, are the ancestors of our culture. In these we find men grappling with problems not very different from our own – questions of political power, military conquest and defence, the administration of justice, commercial expansion, technological development, religious and secular explanations of the world – and arriving at solutions that have many parallels with today.

But this is not surprising, for men are still men. Mankind has not changed, and an Ancient Egyptian, a Minoan from Crete or a Sumerian from Mesopotamia was neither more foolish nor more intelligent than we are. Social and economic pressures may have altered, public opinion may have favoured different values from those we now accept, but, by and large, they were driven by the same urges and succumbed to the same influences. Knowledge is not the prerogative of modern man, and if we have made more material progress, this is only because we have the advantage of almost fifty centuries of additional experience. Or rather, we should have had if nothing had been lost. As it is, we know that there are breaks in our

knowledge, that some discoveries were forgotten and have had to be rediscovered.

That there are gaps is clear from research in the last century and in more recent times. For when Austin Layard uncovered the remains of Nineveh and other Assyrian cities a mere 130 years ago, virtually nothing was known of this vast civilization except for piecemeal and inadequate biblical references. Yet at Nineveh there was the royal library of Ashurbanipal dating from about 750 BC, hidden until Layard found it. And at Nippur, south of Babylon, another large library was unearthed only eighty years ago, a library of clay tablets a millennium older than Ashurbanipal's. These and other Mesopotamian finds have disgorged an immense wealth of early mathematical, scientific and technological tablets which, once their cuneiform text was deciphered, have shown us an advanced and enlightened civilization that was already highly developed when Abraham left Ur and began his nomadic trek across Iraq and Jordan.

But it is not only in the long-forgotten records of Mesopotamian civilization that discoveries lost by later generations may be found. In Ancient Egypt there was a wealth of information, most of it practical, hidden away in tombs that were sealed and left severely alone by all except the tomb robber who was interested only in the material value of the items he stole and was blind to the intellectual significance of his loot. Yet after the invention of papyrus in the third millennium BC, some cultural evidence was committed to writing, and a portion of this filtered to the West, partly through semitic sources, partly by way of Crete and other Mediterranean islands.

Today, Ancient Egypt is popularly remembered for its early monuments – the pyramids, the Sphinx, the temple at Heliopolis – for its funerary rites and the mysteries of its religion. Yet there was a later period which is just as significant to the scientist and technologist. This is the time when Greek culture was prized in Egypt, the time when the dynasties of the Pharaohs gave way to the rule of the Ptolemies. In the latter part of the fourth century BC, Alexander the Great made his vast conquest of western Asia, controlling territory stretching from his own land of Macedonia in the west to Pakistan in the east. But in 323 BC – as soon as he was dead – disintegration set in, his generals quarrelling among themselves and dividing up the spoils.

So it was that Egypt came under the control of Ptolemy Soter, who made his capital at Alexandria and established there not only his dynasty but a museum and library that became a magnet for Greek philosophers. How large the Alexandrian Library was, how extensive the Museum may have been, no one can now be certain. There are accounts that say it contained 100,000 papyrus rolls, or 200,000, or even half a million, although this last figure probably refers to the number of works rather than the number of separate rolls. At all events, it was the largest collection of written material in

8

the ancient world, and if it had remained unscathed by time it would have been a treasure house of the accumulated wisdom not only of Egypt and the entire Greek civilization, but even of those who lived beyond the boundaries of the Mediterranean basin. Its catalogue and the breadth of its collection were immense, and to swell it further Ptolemy III decreed that every visitor to Alexandria must surrender his books to the Library where, if not already in the collection, they were kept. For this enforced generosity the donors received a cheap papyrus copy.

But what started out as so magnificent an enterprise began a gentle decline after 145 BC, though when Julius Caesar laid siege to Alexandria a century later, it was still an extraordinarily rich collection. Certainly some books were burned during the siege, yet the Romans, who thought of themselves as the true liberators of Egypt, seem to have left the Alexandrian Library and Museum largely intact. They may have seized some manuscripts to send to Rome, but the tragic wholesale destruction of the Library did not come until late in the fourth century AD. Then the despoilers were not the Romans but the Christians.

On the advice of the fanatical Theophilos, bishop of Alexandria,

who detested all pagan learning with pathological vehemence, the emperor Theodosios ordered the Library to be burned. Although some books were secretly carried to safety, by 416 AD the destruction was to a great extent complete. And what little remained was finally demolished two centuries later by the Muslims under the Caliph Omar, who used as his reason that either the material still there was already in the Koran – in which case it was superfluous – or it was not and must, therefore, be heretical!

Fortunately, although the Alexandrian Library was the best equipped in the ancient world, it was not the only one. Large libraries existed at Antioch, Pergamum and Rome, while smaller collections of material were scattered all over the Mediterranean countries. To be sure, Pergamum has not survived, and others did not pass unscathed through wars and natural disasters. But when the Muslims consolidated their conquests in the eighth century AD and settled down to cultivate the arts of peace, there was still a wealth of written knowledge. At Baghdad, Haroun-al-Rashid, the sultan immortalized in the *Arabian Nights*, enthusiastically set about collecting scholars – Muslim, Christian, Jewish, African – and feeding them with manuscripts to translate into Arabic, and his example was soon

The giant pharos (lighthouse) at the harbour of Alexandria, shown here in an eighteenth-century reconstruction. Behind it stands the city, which contained the famous museum and library, home of the greatest collection of writings in the ancient world. But by the early part of the fifth century AD, most of this collection had been destroyed.

followed by other Arabian princes. Gradually and laboriously the Muslim world built up an extensive collection of Greek learning, adding to it some contributions of their own. It was this corpus of knowledge that, after a delay of almost 300 years, began to filter into the West.

To begin with, gaining access to Greek learning and scraps of information about the Mesopotamians and Egyptians was a hazardous business. But when Toledo, the western stronghold of the Muslims, fell in 1085 to Alfonso of Spain, scholars could at last journey to a

Christian centre and set about gathering knowledge openly, though the texts they wanted were not in the Greek that they knew, but in the Arabic that they did not. A fresh language and a new script had to be learned before they could start work. And when at last their Latin translations were made, they were mostly only translations of translations, coloured by Muslim emphasis; manuscripts in the original Greek were rare, and those they did find were often commentaries or, worse still, spurious works fathered on the Greek philosophers.

Only slowly was an effective critical apparatus developed to deal

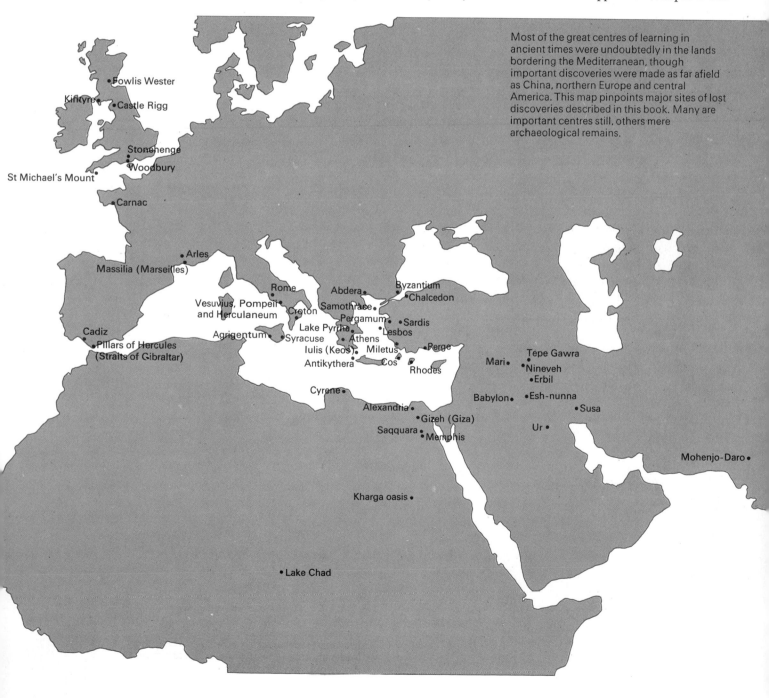

Most of the great centres of learning in ancient times were undoubtedly in the lands bordering the Mediterranean, though important discoveries were made as far afield as China, northern Europe and central America. This map pinpoints major sites of lost discoveries described in this book. Many are important centres still, others mere archaeological remains.

with frauds, and only gradually did original texts come to light; even now there are great unfilled gaps. Yet occasionally papyri still appear, scrolls from the Dead Sea are uncovered, potsherds and clay tablets from the sands of the eastern Mediterranean countries come on to the market to supply tenuous links. Archeological digs and deep-sea salvage help with other clues, and painfully, slowly, another piece is fitted in to the mosaic of ancient culture.

But man travelled much farther afield than is often realized, and learning frequently flourished far from the Mediterranean basin. Nowadays we are more sensitive to other civilizations besides those of classical antiquity: The West African bronzes of Benin show a combination of artistic beauty and technological skill that is utterly breathtaking, the voyage of the Kon-Tiki a boldness and competence in navigational exploration that our forefathers would never have believed possible in a 'savage,' while megalithic structures like Stonehenge display a standard of mathematics and astronomy that speaks for a thriving culture more than a millennium before Caesar invaded Gaul and Britain. Yet this is all comparatively new; awakening to the achievements of world culture is a twentieth-century privilege.

To modern man the past no longer presents a simple picture of continuous growth of Mediterranean civilizations that flourished alone and, in their turn, impregnated his own with their cultural heritage. Studies of the history of science and the development of technology have broadened our definition of what constitutes a civilization; no longer can we be restricted to the arts or the humanities. And once-buried evidence of early time shows that we are not the inheritors of just a straightforward system, simpler than our own. This is too naïve a view. Other cultures had qualities and complexities not very different from those we experience today; so much is obvious even though what we know of them is far from complete. They experienced influences and developments unknown and unsuspected until recent times. Above all they contained discoveries that were made and then lost – lost because they were only partially understood, because they flew too strongly in the face of received opinion, or even because they outpaced the technical knowledge of their own time.

But it has always been so, and we should not be surprised that human ingenuity frequently hit on ideas that are often considered purely modern, ideas that cover the whole gamut of human endeavour – from simple natural history and studies of the human body to the most grandiose concepts about the universe, from down-to earth matters of commerce to the abstracts of pure mathematics, from primitive ways of raising water to mass production techniques, and all a millennium and more before the Renaissance. These are the lost discoveries – the product of man's curiosity – that might, perhaps, have transformed our civilization.

The magnificently painted ceiling of the tomb of Rameses VI, king of Egypt some 3,000 years ago, shows the Egyptian constellations of stars. Egyptian astronomers were far less precise than those of Mesopotamia, and their constellations covered vast stretches of the heavens. Modern astronomers have been able to identify only a few. This photograph is by Professor Owen Gingerich.

CHAPTER 1

Star-Searchers and Mystics

FAR back in antiquity, it was the heavens that offered the greatest scope for intellectual adventure, for exploration by the emerging mind of man. In the clear, unpolluted skies of the ancient world – in the Mediterranean countries, in China, and in the Mexican isthmus – the stars shone forth with a brilliance that commanded attention. There were no street lights to blind the eye, and precious little artificial lighting at home to mar the majestic pageant of the night sky. The Moon and stars ruled the hours of darkness as the Sun did the light of day. Thus it was natural for man to look at the skies and speculate about what he saw, although then, as now, the detailed study of astronomy was the province of the philosopher and the scientist, the man with a yearning to seek the truth about his place in nature.

Even a cursory glance at the sky showed that some stars were bright, others dim, and that all moved across the sky, silently, remorselessly, from dusk to dawn. And at times there was also the Moon, shining down with a brilliance that lit the countryside brightly enough to make it possible to walk about in safety. Observation showed, too, that the Moon changed shape as the weeks passed. Sometimes it appeared as a thin crescent in the evening sky, sometimes a round, full, brilliant globe in the midnight sky, and then it would slowly change back to a crescent that waned even thinner before the approaching dawn. And the Moon repeated this cycle month in and month out, without faltering – regularly, relentlessly, irrespective of the fate of ordinary men and women.

The astronomer also noticed another simple but nevertheless important fact about the Moon – it moved among the stars. So not only did its shape alter as the days passed, but it also changed its place against the backcloth of scintillating pin-points that appeared in the same patterns night after night. In brief, the Moon had a wandering motion of its own, besides regularly rising and setting like the rest of the stars.

But the Moon was not the only wanderer in this night sky. Meticulous and extended examination of the heavens revealed other errant bodies – stars that were not fixed like the rest. The most noticeable were the morning and evening stars, which shone brilliantly at dawn and dusk, although the remaining wanderers were also bright, even if they could not quite match the other two. Astronomers recognized five of these bodies besides the Sun and Moon, and as soon as they really began to examine them, they found themselves faced with a serious problem. The Sun and Moon moved along straightforwardly enough, tracing direct paths across the patterned background of the sky, but the five wandering stars did no such thing. Although they moved forwards most of the time, they seemed to have unaccountable lapses, sometimes standing still, then moving backwards for a time until, once again, after a second short pause, they took up their appointed course. Here, then, was a philosophical and scientific puzzle. Why did the wanderers behave in this way, and

what explanation could be found to account for the observations?

There were two requisites if the right explanation was to be given. First, the starry background had to be precisely observed, the stars themselves organized into patterns, and their positions plotted. Second, the movements of the Sun, the Moon, and the wandering stars – planets as we now call them – had to be noted down as well. Only when these two demands had been met could the astronomers of the ancient world hope to find an accurate explanation for the motions. Of course, we are to some extent looking at this with hindsight; explanations were made before the planetary motions were known at all precisely, and it is doubtful if anyone sat down and decided on a specific plan of research to solve the riddle. Yet, as time went on, this was, in fact, what happened.

So far as dividing the sky into constellations, the Ancient Egyptians tried their hand at this, but breaking it up into huge figures, each of which covered a vast area of sky, so that their constellations were so big as to be unmanageable. In fact they were so vaguely defined that although a few, like the constellation of the hippopotamus, have been recognized by contemporary scholars, there is still doubt about how the rest should be identified with our present-day reckoning. This aspect of Egyptian astronomical laxity throws into relief the differing casts of mind that have occurred throughout man's intellectual progress, because while the Egyptians were satisfied with their unwieldy celestial figures, the Sumerians, their near neighbours in both time and place, were applying their flair for mathematics to the astronomical conundrum.

The Sumerians were the race that populated the fertile valleys of the Euphrates and Tigris in Mesopotamia (present-day Iraq). These people and their successors, the Akkadians and later the Babylonians, worked with greater precision because they had a highly developed form of arithmetic. Their culture lasted from the third millennium BC, when the Sumerians themselves ruled the country, on through subsequent conquests. Intellectually they conquered their conquerors, and the mathematical tradition permeated first Babylonia, and continued to the end of the pre-Christian era.

Clay tablets still exist from way back in the second millennium BC, by when the Akkadian power had waned, and soon after the famous King Hammurabi had drawn up his legal code that pre-dated the Mosaic law by some two centuries. They give us an insight into Babylonian thinking about the heavens which must have been the result of a long tradition of mathematical analysis in astronomy. To be specific, the cuneiform texts make evident a quite sophisticated system in which the entire sky was divided up into three zones, each of twelve sectors. The constellations were mentioned, and the planets too, with a number sequence being given in each case. It appears that what the Babylonian astronomers were doing was computing a series of numbers, running from small to large and back again, to express the changing planetary positions. This was a truly scientific

A guide-book for ancient star-gazers:
This cuneiform tablet, dating from the Seleucid
period of Babylonian history in about the
fourth century BC, is an ephemeris of the
planet Saturn. In the form of tables of numbers,
it predicts the wandering motion of the planet
against the back-cloth of the 'fixed' stars. It is
a form of astronomical almanac, and nothing
contrasts so clearly the precise Babylonian
mind with that of the Ancient Egyptians.

approach to the problem of determining where any planet would be at a later date.

If the numbers are plotted on a graph – an exercise that is not necessary to bother with here but which the modern scientist finds illuminating – they form a zig-zag pattern, illustrating that the Babylonians had recognized that the planets periodically altered their motions as they wove their paths among the fixed stars. And this zig-zag technique was improved as time went on, so that a famous text of the seventh century BC shows these very numbers used for a rigorous mathematical determination of the complex motion of the Moon. Four centuries later, by the time Mesopotamia was under the rule of the Seleucids, and the Alexandrian Library had been established 800 miles away to the west, tables of future planetary positions based on zig-zag calculations had long been available. They were the equivalent of today's astronomical and nautical almanacs.

These Babylonian methods were unique. The Western world inherited the Greek tradition, which used more cumbrous geometrical techniques rather than arithmetical calculations to predict lunar and planetary positions. Not until the seventeenth century AD

were similar, although more elaborate, arithmetical methods used again, and then not because of any Babylonian tradition, but because they were discovered afresh. The Babylonian development went into cold storage and only in this century has evidence of their genius been recognized.

Computation was not the only aspect of astronomy in which the people of the two rivers excelled; they also had novel and penetrating ideas about the nature of the universe. Cuneiform tablets from the Kassite period – about 1500 BC – show that the Sumerians and Akkadians had been thinking about the distances of the stars, and although their figures were lamentably small by today's standards, they had made a definite step in the right direction. What they did was to conceive of the universe as a series of eight spheres, each one nesting inside another, and each governed by its own deity. This, in itself, was nothing astonishing; it seems to have occurred again and again throughout antiquity and even to have continued right up until the days of Newton. But what is surprising is their concept that the spheres were linked with different constellations.

Using the constellation names we now use – names that have come by way of the Greeks from the Minoan civilization of some four millenia ago – there is a cuneiform text that can be translated: '19 ["miles"] from the Moon to the Pleiades; 17 from the Pleiades to Orion; 14 from Orion to Sirius . . .' and so on for eight stars or constellations. In the night sky the Moon is, self-evidently, the nearest body, and the others were obviously placed at varying distances depending on their apperance.

This universe seems to have been rather small, for the farthermost sphere was only 120 'miles' away, though we do not know what distance they meant by a 'mile'. But this is of no significance. What matters is that at least 3,500 years ago men were thinking of the stars as scattered throughout space. Yet within a millennium this bold idea, this presage of our modern views, was forgotten. The stars became fixed to the inside of a dome or a box; Babylonian, Egyptian and, to begin with, Greek, accepted this concept. The stars all lay at the same

left above An artist's impression shows the universe as the Babylonians conceived it. The solid dome of the heavens is supported by the range of mountains that ring the Earth (which is itself mountainous) and its surrounding sea.
left The kufa was the common form of boat used in Babylonia; perhaps the shape of the Babylonian universe was modelled on that of the upturned kufa's hull.
right A relief shows the Egyptian pharaoh Amenhotep and his family offering gifts to the Sun-god Ra. It clearly illustrates the central importance of the Sun in the Egyptian scheme of the universe. They believed the Sun to be source of the life-force that brought food and life to this world.

famous enough. The first was Philolaos of Croton, a follower of a unique sect that was part scientific and part mystical. The other was Aristarchos of Samos, one of the greatest astronomers the Greeks produced.

Philolaos was born some time in the latter part of the fifth century BC, but there is no evidence to pinpoint his dates any closer; all we do know is that he was a contemporary of Socrates, and seems probably to have been a few years older. Whether he was born in Croton, on the east coast of the toe of Italy, is also uncertain, but there is no doubt that he lived and worked there. Philolaos was a mathematician and an astronomer and, above all, a disciple of Pythagoras.

Pythagoras, now remembered almost solely for his rule about the relationship between the sides of a right-angled triangle, was not only a mathematician but also a mystic, and founded a sect known far and wide as the Pythagoreans. He believed the world, its meaning and the divine harmony he thought he saw in it, could all be expressed in numbers, and for a time he preached his ideas at huge public meetings. These were so popular that women even broke the conventions of society and flocked to hear him. His sect, which also taught belief in the transmigration of souls, was exclusive; members wore a particular habit and went barefoot. In brief, Pythagoras was a strange mixture of a man, half mathematician and half religious leader, a cross – as Bertrand Russell once put it – between Einstein and Mrs Eddy.

Astronomically, Pythagoras – and therefore Philolaos – accepted two basic ideas: one that the universe was spherical, the other that the

left The ancient Greek idea of the universe placed the Earth – flattish and surrounded by sea – at the very centre, with a complete sphere of the heavens surrounding it. The Sun, Moon and all the planets rotated around the Earth – a false idea that was to maintain its grip on men's minds for more than two millennia.

right Even as late as the sixteenth century AD, the Danish astronomer Tycho Brahe was influenced by the Greek system. His scheme placed the Earth at the centre of the universe. The planets moved around the Sun, which in turn moved around the Earth.

Sun, Moon and planets all moved in circular orbits around a stationary Earth. The concept of a spherical universe instead of the hemispherical dome of the Babylonians and Egyptians had arrived – if we except the then forgotten spheres of the Sumerian-Akkadian philosophers – some time around the eighth century BC. Its source is uncertain, but evidence for it is found in Homer's *Iliad*. The idea caught on, not least because a sphere is a more elegant shape than a hemisphere, and the Greeks had a love of form.

The idea of circular orbits for the planets was due to Pythagoras – or so his disciples claimed, although, to be frank, they always fathered the best ideas on their leader so that it is now impossible to know precisely what he did, and what was really the work of others. No matter, conceptually it was an aesthetically pleasing Pythagorean

notion. A circular orbit has neither beginning nor end, and so befits the motion of celestial bodies that, after all, appear ageless and eternal – a belief that was not confined to Pythagoras and his sect. Again, this motion was regular: it never changed but continued at precisely the same rate all the time. This was just the kind of motion that one should expect with heavenly bodies, which should not debase themselves by ignominiously hurrying here and dragging their feet there – or so the Pythagoreans taught.

It was a view that was to permeate astronomy for the next 2,000 years and more, and was to play havoc with Greek explanations of planetary motion in the next seven centuries. It was one belief whose loss might have been of signal benefit to later generations but, with the perversity that sometimes seems to dog mankind, it became

enshrined into an unforgotten principle. The Earth was the centre of the universe.

Philolaos inherited these opinions, but he extended the ideas. At the time, astronomers – Pythagoreans and non-Pythagoreans – knew of the existence of seven celestial bodies that wove paths among the stars – the Sun, the Moon, Mercury, Venus, Mars, Jupiter and Saturn. These they could see; these they fitted into their planetary schemes. But the Pythagorean order extolled the number ten, and if, as they believed, it was the most significant, the most sacred, of all numbers, then it followed – to them if not to us – that it should be echoed in the universe. There should be ten celestial bodies, not eight. (They counted the rotating sphere of the stars as one complete body.) This was the problem Philolaos set himself to solve, and it is

for his solution that he is remembered today, even though it was forgotten for a time once the Alexandrian records had been destroyed.

What could be done? Nothing less, Philolaos realized, than a complete re-appraisal of the entire scheme of the universe, and this, astonishingly enough, he managed to achieve. In the centre of the universe he placed not the Earth, nor yet the Sun, but a central fire which gave life to the whole Earth and warmed that part of it (the opposite side to the Mediterranean) that the Greeks thought had grown cold. Around this central fire, nine bodies moved in ceaseless circular orbits – the Sun, the Moon, and the five planets, making a total of seven, the sphere of the stars, making eight, the Earth and a body called the 'counter-Earth', which always lay on the opposite side of the central fire from the Earth. This brought the total number

of bodies up to the sacred ten. It was all most convenient, for it offered a simple and straightforward explanation of planetary motions, backwards as well as forwards. In short, it was carefully and brilliantly thought out.

The Philolaic universe was an adventurous step in astronomical speculation. It illustrates the boldness of the Greeks when it came to casting aside cherished scientific opinions and forging new theories. That Philolaos was wrong does not concern us; that he suggested that there was a central fire where he should have put the Sun is of no consequence. What is significant is that he dethroned the Earth from its privileged position in the centre of the universe and turned it into a body that was like the other celestial bodies, moving through space in a regular circular orbit. He also called the Moon a 'heavenly Earth' and so was the first not only to propose a moving Earth, but also to suggest that another celestial body was no different in essence from the globe on which he lived. Not for another 2,000 years was this to be suggested again, although it was only three centuries before a second scheme with a moving Earth came to trouble the calm of accepted opinion. Yet this neither detracts from Philolaos's pioneering achievements, nor makes us regret any the less that his views became lost in the Alexandrian holocaust, almost 800 years later.

Our other figure in Ancient Greece whose work became forgotten was Aristarchos. He was not the only Greek philosopher-scientist of that name – there was another Aristarchos who became a director of the Alexandrian Library and Museum – and the two are sometimes confused. Fortunately they were born in quite different places; the Aristarchos of the Alexandrian Library came from Samothrace in the northern part of the Aegean, and the Aristarchos we are concerned with from the island of Samos, some 250 miles to the southeast. Samos had seen the birth of a number of philosophers, but by the time Aristarchos was born – about 310 BC – this once proud and independent state was part of the Seleucid kingdom. However, it still had strong links with Athens, and it was most probably there that Aristarchos received his scientific training. Where he later worked is uncertain, but again there is the possibility it was Athens. At all events, it is what he achieved that is significant rather than where he did it.

Aristarchos is important to us for two reasons: He made the first estimates of distances of the Sun and Moon with any pretensions to accuracy, and devised the original idea of what we now call the solar system. His measurements were ingenious and based on the geometry

It took the brilliant mind of Copernicus to reveal the true form of the solar system. But even his was really only a re-discovery of the ideas first formulated in Greece by Aristarchos some 1,800 years earlier. This plan – with the planets, among them our Earth, orbiting the Sun – was the greatest of all astronomical lost discoveries.

involved in measuring the angular positions of the Sun and Moon. Essentially what he did was to compute the distances of the Sun and the Moon by considering their relative positions when the Moon appeared cut in half or, as we should now put it, when the Moon reached first quarter. At this time the Sun lies directly to the side of the Moon – a long distance off, certainly, but directly to the side. This is why the Moon appears exactly half illuminated, no more and no less.

To complete the picture, Aristarchos needed to know two other things – the angle in the sky between the Moon and the Sun, and the apparent size of the Moon in the sky. This last quantity is not difficult to determine, but Aristarchos unfortunately used an incorrect value that was four times too large. Only later in his life, after he had written a book about his measurements, did he recompute, using a correct value. As far as the angle between the Sun and Moon was concerned, this is a very difficult angle to measure precisely, and it is perhaps not surprising that Aristarchos got it wrong. All the same, his mistake was not gross, but only a little more than three per cent out. Nevertheless, in his scheme of measurement it was a vital angle; even a mistake as small as three per cent would throw the final figures. When coupled with the immense difficulty of determining just when the Moon's disk appeared exactly half lit, as well as the wrong value for the Moon's apparent size, it is little wonder that he gave distances for the Sun and Moon that were far too small.

But the actual figures are not as important as one might think, because in the third century BC no one had previously considered actually measuring the distances; at best they said what amounted to no more than the simple fact that the Sun and Moon were too far away to touch. Aristarchos was the first to apply mathematical techniques to solve the problem, and this is the significant thing. What is more, he went so far as to develop new mathematical methods to do so. In fact he very nearly hit on the mathematical technique of trigonometry which later generations were to derive from his mathematical treatment of the distance problem.

We are much more concerned here with Aristarchos' methods than with his results. They were taken up and expanded by Hipparchos about 150 years later, and used so successfully that he obtained a distance for the Moon that was very nearly correct, although the Sun's distance still eluded him. Not until almost 2,000 years later were observing methods precise enough to give the correct value. But a reasonably close estimate could conceivably have arrived earlier if only Aristarchos's work had been widely applied. Unfortunately it was not, and his methods went into the limbo of the lost, until they were again considered at the time of the Renaissance, when so much Greek learning erupted into Western Christendom.

What is true of Aristarchos's mathematical measuring techniques in astronomy is true, too, of his astonishingly independent ideas about planetary motion. For he went further than Philolaos, and did so without having to invent mythical bodies like a counter-Earth and a central fire. In fact he broke entirely with the tradition blessed by Aristotle, and conceded as correct by every philospher-scientist of his time. The Earth, he said, moved in space. And this is not all, for he claimed that it also spun on its axis, so that the sphere of the stars did not spin round once every day; their motion was merely apparent and due, as we now know to be a fact, to the Earth's daily rotation. This was a breathtaking idea – it played havoc with the whole body of Greek physics – yet it did provide a simple and elegant explanation of the movements of the planets and of the Sun and Moon.

Aristarchos placed the Sun at the centre of his scheme, and considered the planets to be in orbit around it. Like Philolaos, he dethroned the Earth from any special and privileged position and made it one with the other five planets. Yet unlike that of Philolaos, his scheme was not mystical; it was a purely mathematical theory. Certainly it may have owed something to Philolaos and something, too, to Heracleides of Pontus who, two generations earlier, had evolved a hybrid planetary system. But let us make no mistake: Aristarchos's proposals formed the first thoroughgoing heliocentric theory and pre-dated that of Copernicus by 1,800 years.

When, in AD 1543, the Copernican theory was published, a storm of disbelief and argument arose; only gradually did it take a hold on the scientific community. Yet in spite of some highly organized opposition it was an accepted part of the astronomical universe within a hundred years, and revolutionized the entire outlook of man. And not only was this so astronomically. The revolu-

An apocryphal portrait of Pythagoras from the *Nuremberg Chronicle*, published in 1493. Mathematician and religious mystic, he had strong ideas on the harmony of the universe, believing that this could be expressed in numbers.

tion touched every aspect of intellectual life because, as with Aristarchos, it removed the Earth from a preferential place in the universe and relegated it to the role of a satellite that orbited the Sun. Man was no longer the special favourite of creation. What then might have happened if the Greeks had paid heed to Aristarchos instead of accusing him of impiety? What complete revaluation of man's place in the universe might there have been if this astonishing theory had been taken up? Unhappily, it was forgotten. Yet, seeing what occurred after Copernicus, it is clear that a great slice of human intellectual history could have been changed out of all recognition.

Also from the *Nuremberg Chronicle*, a supposed portrait of Aristarchos of Samos, father of the modern view of the solar system. Such fifteenth-century portraits were designed to convey the personality rather than the likeness of the person concerned.

CHAPTER 2

Calender-Priests and Clockmakers

As technological development has grown, so the measurement of time has become ever more critical, until, in highly organized societies, people are ruled by the clock. The era of the mechanical clock has made us believe that time has an independent existence, and that we can split it into precise divisions of hours, minutes, seconds, to make the timetable the pattern for our day. We conceive of time as flowing past us, blind and relentless, but in the ancient world the outlook was quite different. Time was cyclic, a process that repeated itself, a perpetual roundabout of seasons. And daily life was not so closely governed by time; as the days lengthened in summer, the hours stretched with them, and shrunk when winter brought a shorter day. Time was a local phenomenon; there was no link between noon in Babylon and noon in Alexandria, nor yet between the weeks and months they used. Each city, state and principality had an independence of time-reckoning that would be anathema to the international world of today.

Yet there was one over-ruling guide to time, one aspect that was common to all races and all classes of men: the slow cyclic pageant of the heavens. The rising and setting Sun, the ever-changing face of the Moon, the appearance and re-appearance of bright stars and planets – all these formed a celestial clock that pointed the way to the timekeepers of antiquity.

There have always existed two basic pressures to make man aware of time, at least since he ceased being purely nomadic: the need to sow and reap, and the need to order the religious festivals and fasts that give punctuation to everyday life. But these led him into difficulties very early on; all too soon he found himself face to face with incompatible, incommensurable quantities. His difficulties came because he used both the Sun and the Moon as his primary guides, and they move through the heavens at quite different rates. How does one reconcile the Sun's movements against the backcloth of the stars, which has a period of just under $365\frac{1}{4}$ days, with the Moon's $29\frac{1}{2}$-day cycle? The nightmare that confronted every early calendar maker was how to combine a cycle of religious festivals determined by the Moon's phases with an agricultural calendar based on the Sun.

Sometimes this was solved by a strange combination of natural circumstances, as in the case of the Yami fishermen of the island of Botel-Tobago, near Formosa. They use a lunar calendar, and sometime about March – the precise date depends on how far out of step with a seasonal (solar) calendar it happens to be – they go out in their boats on the night of the new moon carrying lighted flares to attract

Time has always been important to man, and the oldest method of measurement is to follow the Sun's movement through the sky by watching its shadow. These tribesmen in Borneo are using a gnomon, or shadow-stick – a method that dates back far into prehistoric times.

the flying fish. If the fish appear, their calendar is correct and it is time for a new fishing season. However, if the lunar calendar sends them on too early an expedition, no flying fish are to be found, and the Yami wait until the next new moon. In this purely empirical way, they have over the centuries kept their calendar in step with the solar year by adding these extra lunations (lunar months). But the Yami are lucky; such a combination of natural phenomena is rare, and the would-be compilers of calendars have usually to adopt more prosaic – and more tedious – methods.

An agricultural calendar calls for a division of the year into seasons, and this brings us straight to the problem of measuring the Sun's position in the sky. One simple solution is to use a pole stuck vertically in the ground. At midsummer, when the Sun is at its highest, the shadow this gnomon casts is at its shortest, and at midwinter the shadow reaches its greatest length. This is a simple, primitive, but effective method, and the technique can be extended to define the times in between.

But there is another, more sophisticated way of determining the Sun's position; this is to track its path across the background of the fixed stars. To do so presumes a knowledge that the stars are still in the sky when the Sun is there, even though they cannot be seen, but this can be expected from any civilization like those of Babylon and Egypt which had developed enough to group the stars into constellations. To find the whereabouts of the Sun is not as difficult as it sounds, because once the sphere of constellations – or, more particularly, the Zodiac, the band of constellations across which both Sun and Moon pass – is charted, then, if the constellations visible just after sunset and just before sunrise are observed, it is clear that the Sun must lie in between. This was the method devised and used by both Babylonians and Egyptians. But surprisingly, in view of their general lack of a precise astronomy, it was the Egyptians and not the Babylonians who made best use of it. This may well have been due to the practical bias that coloured the Egyptian character, but whether or not this is the reason, the fact remains that they computed a civil calendar that was simple and effective – so simple and effective, in fact, that there are people today who wish that we would re-adopt it.

Part of the Egyptian success was due to local conditions, for their whole agricultural system was based on the annual inundation of the Nile. This was the essential fact of nature that ruled their entire economy; this and its aftermath were what made the country habitable. So the Egyptians needed, above all, a calendar pegged to the Nile floods. To achieve this they recognized three seasons, each of four months: flood time, seed time and harvest. The first of these, the time of the flood, was presaged by the rising, just before dawn, of the very bright star Sothis (our Sirius), when the Sun, as we should say, was in the constellation Gemini. This led them to begin their day at dawn – straight after Sothis had risen – and their months with the appearance of the thin crescent of the new Moon.

below The great Aztec sunstone, or calendar stone, 12 feet in diameter, is a recording carved in stone of the cyclic movements of heavenly bodies. It enabled the Aztecs of Mexico to regulate their calendar and predict such events as eclipses with great accuracy.

223 lunations (i.e. cycles of the Moon's phases) = 18 year eclipse cycle.

235 lunations = 19 years = 1 period of repetition of the phases of the Moon on the same days of the year.

1236 lunations = 100 years of 365 days.

390 lunations (upper rectangle) + 390 lunations (lower rectangle) = 780 lunations = the same number (780) as there are days in a synodic 'year' of the planet Mars (i.e. the same number of days as it takes for Mars to move right round the heavens as seen from the Earth).

A total of 585 lunations = almost the same number (584) as there are days in a synodic 'year' of the planet Venus.

The Caracol or observatory at Chichen Itza in Mexico dates from Mayan times, and shows another method of watching and timing star movements. There are alignments along passages cut in the walls, so that an observer inside can plot the position of a star or the Sun or Moon against the edge of a passage.

Using the heavens as clock and calendar demands accurate surveying of the positions of heavenly bodies. This artist's reconstruction shows how Ancient Egyptian astronomers used the merkhet – an instrument consisting of a plumb line and a forked stick – to find an accurate north-south line and, by timing the moment a star crossed this meridian, map its position in the sky.

Some time far in the distant past – perhaps as early as the fourth millennium BC – when the crescent Moon and the early morning rising of Sothis coincided with the annual inundation, the reckoning began. At first it was purely lunar. Gradually, as the years passed, the lack of coincidence between the appearance of the crescent Moon and the rising of Sothis would have become a tedious nuisance. The same would apply to the fluctuations in the times of the inundations, which were dependent on the melting snows in the far-off mountains, and so did not follow each other with clockwork regularity. The two constant factors were the unerring regularity of the Sun's motion in the sky, and the pre-dawn rising of Sothis. Careful study showed the Egyptians that every twelve months plus an additional eleven days, Sothis again rose as he should. Nothing, absolutely

nothing, altered this. So the priesthood, one of whose tasks it was to determine feast days and thus regulate the calendar, decided that the lunar reckoning should be brought into step with the solar/stellar one by adding an extra month every two or three years, whichever was necessary.

This insertion of an additional month may have been a mystery that the priesthood kept to themselves, but as economic life became more highly organized, the use of an irregular-length year became unpopular. There was a general demand for a settled civil calendar that varied far less, a calendar in which an additional period as long as a month did not suddenly appear, even by premeditation or design. An examination of past records showed that the length of a year was close to 365 days, and this was the average year the Egyp-

tians adopted. They then instituted an average value of 30 days for a month – a little long, but a satisfactory compromise with the $29\frac{1}{2}$ days that a lunation really took, since four months of 30 days each gave one-third of a year, and so fitted in conveniently with the three Egyptian seasons. There were still five days to be accounted for (12 months of 30 days giving no more than 360 days) and these were inserted at the beginning of the year. For civil purposes an addition of five days regularly, year in and year out, was not disturbing, and far more preferable than the insertion of an extra month every few years.

The 365-day calendar was not exact. On the average, Sothis rises before dawn at intervals of just over $365\frac{1}{4}$ days, and the Egyptians were well aware of this, as of the fact that their civil calendar gradually progressed through the seasons as time went on. But they knew, too, that it would get into step again after 1,461 years – because 1,461 multiplied by 365 makes 533,265, which is the same as 1,460 multiplied by $365\frac{1}{4}$ – a period that was called the Sothic cycle for obvious reasons. For purely practical convenience, then, they decided that it was best to have a fixed civil calendar, even though it slowly fell out of step. The Egyptians were pragmatists. The priests still had a lunar calendar for religious festivals, and they made adjustments to

this when it became too far out with the seasons, as it frequently did.

Another surprising thing is that the Egyptian civil calendar was in use early in the third millennium BC. This pre-dates by some 3,000 years those in use south of the Gulf of Mexico by the Aztecs and Mayans. These seem to have been primarily secular solutions, like the Egyptian, for the Aztec religious calendar – the 'tzolkin' – bore no relationship either to the Moon or to the seasons. It was in fact not celestial in any sense, but was purely numerical, being based on a combination of the numbers 1 to 13 taken cyclically, married to symbolic figures each representing 20 days ($13 \times 20 = 260$). But the practical Aztec calendar contained 365 days, while the Mayan was even more accurate, since it added an extra day every four years, forming a leap-year. Thus they came to the same solution as the Romans, but at least 500 years earlier.

The Mayan calendar was unknown in the Old World: a discovery made, yet lost, because the Mexican isthmus remained unexplored until the Spanish invasion in the mid-sixteenth century AD. And as far as the Egyptian civil calendar is concerned, this too was forgotten. The Western world received its calendar tradition from the Romans, and from the lunar calendar of the Old Testament. The

top An eclipse of the Sun is an awe-inspiring sight, and anyone in ancient times who could predict such a happening would have had very great power.
bottom Stone circles, like this one at Pontypridd in south Wales, could have been used for this purpose. The stones were set out with extreme accuracy by Stone Age astronomers, and were used to observe the Sun's and Moon's most southerly and northerly points of rising and setting. This not only gave megalithic man a precise practical calendar, but also a method of predicting the occurrence of eclipses.

idea of a civil calendar that overruled the religious, and even paid attention to the seasons, was never contemplated. Instead the Western calendar makers, like the Babylonians, Greeks and Romans, became obsessed with cyclic time, and worked out eleborate tables of relationships to link the ever-recurring short lunar periods with the long seasonal years of the Sun. We still suffer from this today in the determination of Easter, although the merits of fixed dating, as the Egyptians used, have at least been recognized in our fixed date for Christmas and the habit of pegging other holidays by calendar dates. We have rediscovered the wisdom of the Egyptians in putting a priority on practical calendar dating – but some 4,000 years late.

The ancient world found its concept of cyclic time echoed and re-echoed in the heavens. The Sun and Moon, the five planets, all went through their appointed courses, returning time and again to the same configurations. Equally impressive in the celestial order were the strange and terrifying phenomena of eclipses – rare, but following some long-term pattern all the same. Perhaps we tend to underplay them today. Eclipses of the Moon, when the full disk turns blood-red, pass by unnoticed in the glare of city lights. And total eclipses of the Sun happen so rarely in the same place – once in 360 years on the average – that most people only hear reports of them, or see a representation on television that is no substitute for one of the most majestic, most chilling of all natural spectacles. But this was not so in ancient times. Eclipses were outstanding events, divinely ordained and presaging, perhaps, godly displeasure; they could only be ignored at man's peril.

And so a vital need arose: the need to tell when eclipses would occur, to predict them so that the priesthood, at least, was not caught unawares. But because eclipses are infrequent, it was no easy task; stories that the Babylonians could predict solar eclipses, and that the seventh-century Greek philosopher Thales learned their secret, are now said to be apocryphal. Yet recent studies of north European megalithic structures like Stonehenge show that, in the second millennium BC, megalithic man was certainly busy with this very problem. Here a whole corpus of important knowledge, enshrined in stone rather than on clay or papyrus, is only now beginning to be understood. In the light of it, perhaps, the Babylonian contribution will have to be re-assessed if we are not to do them an injustice and think of them as inferior to the tribes of northern Europe. But, reassessment or not, there are enough lost discoveries from the second and late third millennia BC to make us stand in amazement at a culture which, a century and a half ago, no one ever dreamed existed in so advanced a state.

Why did megalithic man build in stone? Why did he construct circles of astronomical markers – for that is what the stones are – rather than use simpler sighting instruments? How did he acquire a knowledge sufficiently elaborate to predict eclipses, as well as determine celestial positions with great precision? The answers seem straightforward enough. Stone was used because it weathered well in the cold, damp environment of north-western Europe. Accurate, manageable instruments in metal would have called for an elaborate technology that was not to hand, and any built in wood could not have remained in adjustment for long enough. And, of course, the very nature of stone meant that the instruments had to be large and, to be accurate, they had to cover large areas of ground. In fact, it was the immense spaces between the markers that allowed these men to make the precise observations they did. For their information they must have drawn on previous knowledge, but as eclipses occur so infrequently, this knowledge would need to have been built up slowly over long periods of time – over generations rather than during a single lifetime, over centuries rather than years. The megaliths that are now being studied therefore represent the culmination of a culture that must extend far back into a past, probably as remote as any Mediterranean civilization.

Why did the megalithic men feel a need to know answers to problems like the motion of the Moon in its orbit, or the occurrence of eclipses? Doubtless for the same reasons as the Babylonians and Egyptians – because they wanted to know the temper of their gods, and because of the more prosaic but nevertheless important need of a calendar. That they did not use the pre-dawn risings and post-sunset settings like their Mediterranean counterparts was a force of circumstance, due to the high latitude in which they lived. In the lower latitudes of Egypt and Babylonia, closer to the Equator, the Sun, Moon and stars move from the horizon at steep angles and, in the clear dry skies, it is not difficult to determine exactly where and when they rise and set. But this is far from true for the higher latitudes; the stars move at oblique angles and, in the frequently mistier skies at dawn and dusk, the disappearance or reappearance is hard to pinpoint. Their precise location as they emerge or are swallowed up below the Earth is impossible to determine accurately.

What the inhabitants of these cloudy regions needed to find the alignments were distant markers lined up with markers near by, and this is just what megalithic man constructed with his stone blocks and apparently, in a few cases, with large wood structures, perhaps like Woodbury. This is certainly true of Stonehenge, and it is true of a host of other observing sites. As Stonehenge is a rather complex example, it will be best to look first at two simpler arrangements.

At Fowlis Wester, near Crieff in Scotland, there are two elliptical stone circles, one some twenty to twenty-five feet across, the other somewhat smaller. They are separated by about eight feet. Outside the circles lie large menhirs (tall, upright marker stones), and there is also a rocky horizon with which observing lines using a stone from one of the circles can be lined up for observing the settings and risings of the Sun and Moon. At another site, Castle Rigg, close to Keswick in the north-west of England, there is an elaborate circle, flattened on one side, composed of forty stones. Its diameter is over

above An artist's drawing of an eclipse of the Moon shows the rosy colour of the eclipsed part, due to sunlight being bent and filtered by the Earth's atmosphere. A lunar eclipse can be seen far more readily than a solar one, and although it is not such an awe-inspiring sight, it would have seemed an event of great importance in the lives of unsophisticated peoples.

above Sunrise over the heelstone at Stonehenge and **left** a plan of the site. This, the most complex of all the stone circles in the British Isles, has been termed a Stone-Age computer. It was certainly a vast observatory, and was used to gather calendar information and astronomical data with great accuracy.

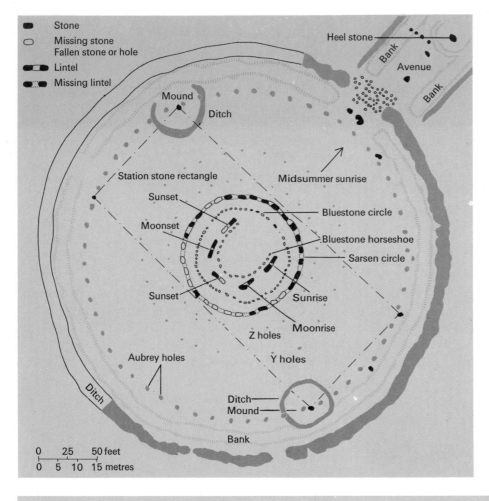

100 feet, and within the circle is a rectangle of ten stones. This was used in a slightly different way from the Fowlis Wester circles, because there are no horizon points with which to line up. Instead, the observer used a menhir at one point of the circle and a menhir at the far side to give his alignment. He sometimes used one of the menhirs lying outside the main circle – an 'outlier' – as an additional aid.

Now, since it is easy to watch the Sun rise and set – and the Moon as well, for that matter – the need to use an elaborate stone circle may not be clear. But such circles and alignments were necessary because what megalithic man needed to know was when the Sun was at its solstice points – that is, its most southerly and most northerly – since it was then midwinter or midsummer, and also when the Moon had reached its most northerly setting position and its most southerly. From this latter information he could compute not only lunations – that was comparatively simple – but also watch the Moon throughout the entire nineteen-year cycle. During this time, the whole orbit of the Moon seems to rotate in the sky. Since the Moon's orbit lies at an angle in the sky, such knowledge would tell the megalith user the times when the Moon would be above the horizon the longest – times that were important when it was the sole effective light at night. But there was also the other reason: the prediction of eclipses.

Eclipses of the Sun occur when the Moon is directly in line with it in the sky and so casts its own shadow on the Earth. This can only take place when new moon coincides with a time when the Moon and Sun lie close to a point where their paths cross. (We are assuming here that the Sun and Moon both orbit the Earth, as megalithic man did, and as the computer does today when calculating the times of sunrise and sunset.) For this reason alone, observations of the most northerly and southerly setting points for the Moon and the Sun were necessary. Then, both orbits, the rotation of the Moon's orbit, and the positions of the Sun and Moon along those orbits could all be found.

Eclipses of the Moon happen when the Sun, Earth and Moon are

East cromlech

| 0 | 100 | 200 | 300 | 400 | 500 feet |
| 0 | | 50 | | 100 | 150 metres |

left Carnac, in Brittany, is the largest known megalithic structure in Europe. Because the sighting points and markers are sited so far apart, astronomical observations could be made at Carnac with extraordinary precision. The main markers were put down in long rows that radiate slightly like a fan, rather than in the normal circular layout found at most other megalithic astronomical observatories.

in line, for at this time the Moon passes into the Earth's shadow. This must be at full moon, but that alone is not sufficient. Once again, the lunar orbit must have rotated so that the Moon is at one intersection of the solar and lunar orbits, and the Sun is at the other. Such an event could only be predicted by megalithic man if observations had been made over a great many nineteen-year cycles.

If this business of observing eclipses sounds a little complicated, that is because it is, and we can only stand in amazement at what erudition and understanding the megalithic civilization showed. And this becomes especially true when we realize that the menhirs were positioned with immense precision – to within a fraction of an inch. This was done by having a common standard of measurement – what one astro-archaeologist calls the 'megalithic yard' and gives as almost $2\frac{3}{4}$ feet in megalithic Britain – and by setting up auxiliary sighting lines using small stones and perhaps wooden posts as well. After all, with menhirs often weighing twenty-five tons or more, and placed on ground that was sometimes swampy, one could not really afford mistakes even of fractions of an inch.

Precisely how these Stone Age astronomers erected the menhirs is uncertain, but probably by first digging a pit with an inclined side, then sliding the menhir into it and, finally, pulling it upright with a pair of wooden shear legs. At Stonehenge there was an additional constructional problem – getting the immense stone lintels on top of the menhirs. Perhaps earthen or wooden ramps were built so that the lintels could be hauled up into position. Yet even if these methods were adopted, they show a high degree of sophistication in mech-anical engineering, especially when we realize that at Stonehenge the stone itself had to come from miles away.

Stonehenge has been likened to a megalithic computer, and in a sense this is what it was. Recent studies and measured surveys of the site show at least six concentric stone circles, although it is possible that further careful investigations may reveal yet more. Not all the circles were built at once. At first there was a ditch, a bank and a circle of fifty-six holes that once contained stones, together with an outlying menhir known as the Heel Stone: this was built about 1850 BC. Then there followed a second ring of stones that seems never to have been completed and, finally, the rest of the structure, including the largest menhirs and lintels, all the work being com-pleted by 1700 BC.

In design, Stonehenge is like the site at Castle Rigg; it works with its own outliers and not using horizon sitings. In many ways it was used in the same manner as Castle Rigg; observations were made of rising and setting points looking across the menhirs and, when appropriate, the outliers. The best-known observation of this type is the one that has received so much publicity in the press and on tele-vision – the rising of the midsummer Sun over the Heel Stone when viewed from inside the menhir circle. The so-called Sarsen Circle – the menhirs with lintels – has thirty stones, and these mark off the changing phases of the Moon, and thus time a lunation, even when it is cloudy and the Moon cannot be seen. The outer ring of fifty-six holes – the 'Aubrey' holes – were used for determining extreme setting points of the Moon, and thus for lunar eclipses, while the main structure also indicated the Sun's risings and settings, which would lead to a determination of the seasonal year.

The megalithic computer, be it at Stonehenge, Castle Rigg, Fowlis Wester, or at any of the other of the 300-odd sites in Britain alone, shows a familiarity with the heavens and an ability to calculate that was lost by the time the Romans invaded Britain in the first century AD. And this is true, too, of the civilization – the same one probably – that built the huge mile-long arrangement of menhirs at Carnac in southern Brittany, in which two egg-shaped circles are connected by long lines of carefully positioned stones, with more distant points and outliers. This computer was elaborate and, like Stonehenge, an advanced structure, but of the people who made it nothing is known. Their engineering standards were high, their astronomy well developed, and yet they have left no other traces that we recognize. Their discoveries were lost until the megaliths they built began to attract attention a century or more ago. Now intensive research is showing how able they were – so able that the more we learn, the more we long to know of their whole culture.

Megalithic circles and the alignment of temples and pyramids in Egypt, Babylonia, and even far away in Mexico, all show the need every civilization felt to observe the sky and prepare calendars, the desire to determine long time periods with increasing accuracy. But what of short periods? How did the ancient peoples divide their day? How could they measure intervals of a few hours or less? In the countries bordering the Mediterranean there was one obvious time-keeper for daytime use, and that was the Sun. Indeed it was so con-venient that it was used in northern Europe too, even though cloud would frequently make it ineffective. Basically the system was to use the Sun to cast a shadow, and to arrange that the shadow fell on a marker or a dial of some kind. Early on the Egyptians used a wooden bar, but in time a graduated dial, cut in stone or painted on a wall, became common practice. It was simple to set up and easy to read. This discovery was not lost; the sundial has an unbroken history that merely shows greater elaboration as the years pass. But it must be mentioned because it coloured the way time was measured until the advent in the West of the purely mechanical clock sometime in the thirteenth century AD.

The Sun remains above the horizon for different periods of time depending on the season. In the summer the days are longest and the Sun will shine for much more than twelve hours, but in the winter it rises later and sets earlier; the days are shorter and the Sun is above the horizon for less than twelve hours. This difference is more marked the higher the latitude; within the Arctic and Antarctic circles it reaches its most extreme, with perpetual day in summer and ever-

Some of the best clepsydrae, or water-clocks, made in the ancient world were the work of Ktesibios of Alexandria. This cutaway reconstruction shows the mechanism of one of these, with its syphon for maintaining a constant head of water and thus a constant running speed. The gear wheels operated the little figure which indicated the time. Note that the drum is marked with unequal hours, so that it could be rotated to indicate longer or shorter hours as the seasons changed.

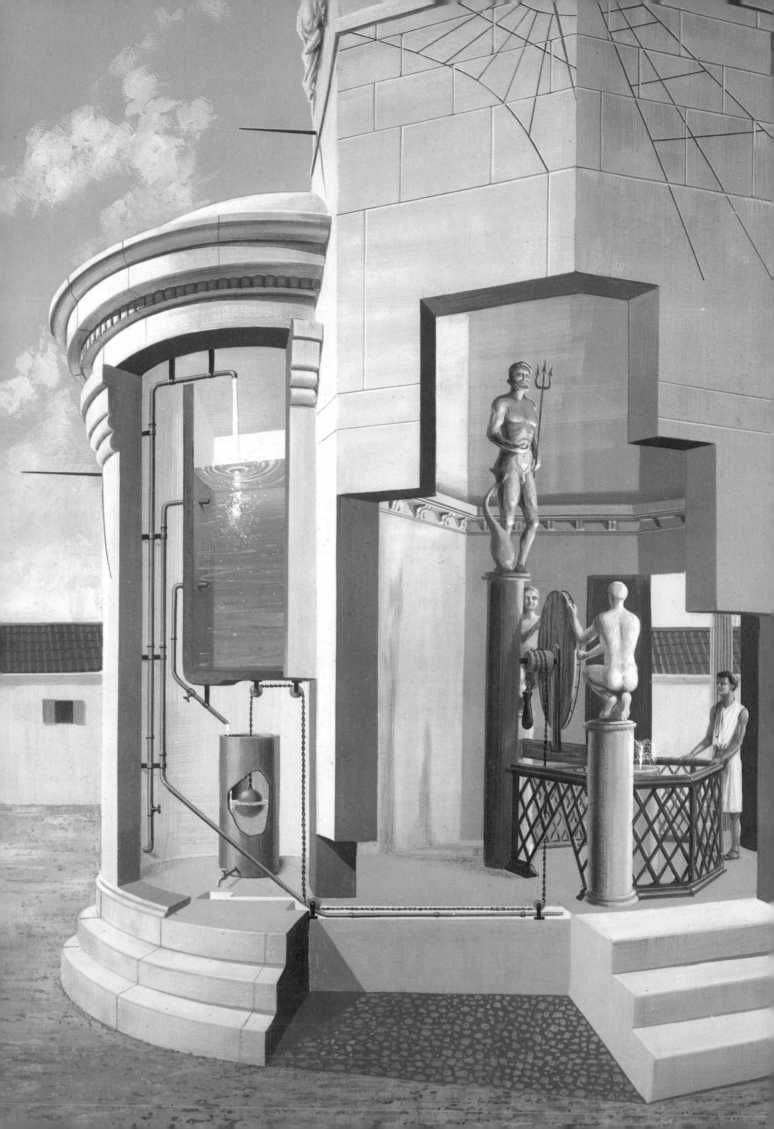

The Tower of the Winds in Athens had a large clepsydra with an elaborate syphon system for maintaining a constant water pressure. This artist's reconstruction is based on the researches of Professor Derek de Solla Price; the actual mechanism has long been lost, although the ruins of the tower remain. It had sundials as well as a public clock.

lasting night in the winter. Since artificial lighting was meagre in the ancient world, the period between dawn and dusk was far more important than now, and it seems generally to have been agreed – at least in the Mediterranean countries – to adopt a fixed number of divisions for each daylight period. The day was divided into twelve hours and the night into twelve hours. But while the number was always the same, the length of the hour varied with the season.

Perhaps this adoption of unequal hours made it difficult for anyone to design a satisfactory mechanical clock, and so postponed the time when they were available. But more probably it was the fact that the sundial seemed satisfactory enough while the tempo of life was less hectic than it was to become after the Renaissance. There were time-keepers that did not depend on the Sun: hour-glasses and clepsydrae that measured the continuous flow of some substance – in these cases, sand and water respectively – and for night used a candle or an oil lamp that indicated the time by its changing level as it burned. In all these instances the unequal hours could be marked so that the timepiece could be turned and the correct seasonal intervals displayed.

Possibly the high-point in the design of a clock along these lines is represented by the clepsydrae built by Ktesibios, who lived in Alexandria about 270 BC and had an immense reputation as an inventor. These were most ingenious devices, since Ktesibios managed to over-come the inherent failing of the clepsydra – the fact that the rate at which the water flowed out changed with time. In other clepsydrae the rate of flow depended on the head of water (or depth above the outlet), and since the head was reduced by the flow, the rate con-tinually grew less. It was like having a mechanical clock that was forever running down. Ktesibios's solution was simple; he had water flow into a cistern which had a siphon. So, provided the inflow was greater than the outflow that registered the time in some way, the level in the clock would always keep constant. He often went to great lengths to construct reliable mechanisms, drilling his holes in gold to prevent verdigris, or in precious stones to guard against wear.

Perhaps his most famous clepsydra was one of a rather different

Su Sung's great clock tower was built at Khaifêng, in Honan province, about AD 1050, and embodied most of the ideas developed by I-Hsing some 300 years earlier. There remains some doubt about the details of the mechanism, but the reconstructions shown here are based on research by Dr Joseph Needham published in volume four of his book *Science and Civilization in China*, and on drawings by John Christiansen.

left The whole tower was some 40 feet high, and the time was indicated both by figures, or 'jacks', passing the windows in the side, and also by bells. The metal sphere made of rings—an 'armillary' sphere —at the top of the building could be used to make astronomical observations. This was driven by clockwork, being fixed at the top of the shaft that carried the gears (**12** and **13**, below) which drove both the jacks and also a celestial globe (**15**, below) that could be examined from inside the building.

type in which a figure with a pointer was mounted on a plunger. As it moved downwards, this plunger forced water out through a pipe at a constant rate. The water flowed into a wheel containing compartments that drove a gear train which, in turn, rotated a column on which the various hours were marked. A seventeenth-century reconstruction of this by the French architect Claude Perrault is illustrated on page 33, since Ktesibios's original design has long since vanished.

A public clock based on Ktesibios's overflow principle was built in the Tower of the Winds in the market-place of Athens. As the drawing shows, this was operated by the outflow of water from a large elevated cistern, the water gradually lifting a hollow metal ball and thus driving a large bronze disk on which lines representing the hours were marked – and probably constellation figures as well. Yet the tradition of public clocks seemed to die with the Greeks. Public sundials remained – the Tower of the Winds had these as well – but

below The whole mechanism was driven by a constant flow of water. Water in the tank **1** was raised by a wheel-and-buckets device **2** to a tank **3**. The wooden cogwheel **4** drove a second wheel-and-buckets arrangement **5** to raise water from tank **3** to tank **6**. From here it flowed into tank **7**. A constant flow of water to the actual 'clockwork' was obtained by a syphon **8**, which led water from tank **7** to the reservoir **9**. The constant flow of water from this reservoir drove the water-wheel **10**. This was the vital controller or escapement of the clock, and is shown in more detail opposite. Via a system of gears **12**, **13** and **14**, it drove the wheels carrying the jacks, the celestial sphere **15** and the armillary sphere on top of the building. Its steady action depended on keeping tank **6** topped up, and this was done by rotating the wheels-and-buckets **2** and **4** by hand. Trough **11** caught the water released by the escapement wheel **10** and returned it to tank **1** to be circulated again.

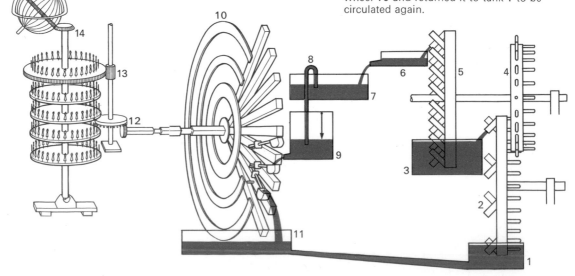

The water-wheel escapement was a brilliant invention. It moved one spoke at a time, once every 24 seconds. As shown here, the arm 1 held the wheel stationary by preventing the spoke 2 from moving clockwise. Water from the dragon's-head spout 3 (leading from the syphon-filled reservoir) poured into the bucket 4. This was fixed to one end of an arm that had a weight 5 on the other end. The middle of the arm was a pivot (as shown for clarity on the arm above, it at 6). As the bucket 4 was filled, a pin on the front caught on the end of the checking piece 7 of the lower balancing lever 8. This kept the bucket in position. But once the weight of the bucket

and water was greater than that of the counterbalance weight 9 at the end of lever 8, this lever swung down. So did the bucket 4 on its pivoted arm, and the pin on the front of the bucket tripped another arm 10. This arm pulled on a chain 11, and thus tilted the upper balancing lever 12 with its carefully adjusted counterweight 13. The lever 12 pulled the link 14 and lifted the arm 1, thus releasing the spoke 2 of the wheel. The bucket moved round to the position previously held by the bucket ahead of it 16. Meanwhile, the next spoke had lifted the upper arm 15 and had moved round to where spoke 2 was. It was

prevented from moving any more than one position because the upper balancing lever 12 swung down again immediately it had tripped, so that the arm 1 caught the next spoke and held it. The whole cycle was now repeated, with water flowing into the next bucket. Meanwhile, the bucket in position 16 that had just tripped poured its water out into the trough below. For clarity, only three arms with buckets have been drawn here; in fact there was one between every spoke.

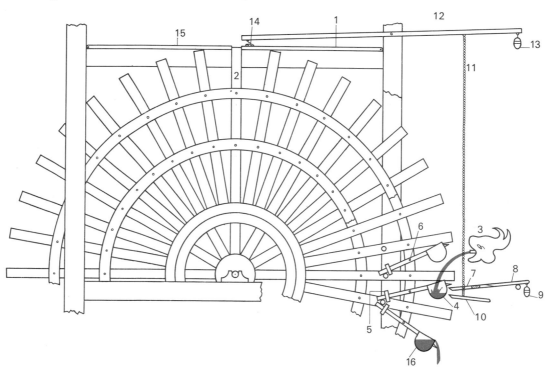

the art of the accurate clepsydra was lost, only to be rediscovered in the Renaissance when the full import of Ktesibios's ingenuity was appreciated. But by then the mechanical clock had overtaken him.

Just how the mechanical clock arrived in the West is still a mystery, but it may be that it was due to rumours of a Chinese invention of 500 years earlier. At all events, it is to a Buddhist monk called I-Hsing and his colleagues that we owe the soul, the essential device of the mechanical clock – the part that later became known as the escapement. Working around AD 723 at the College of All Sages in Chhang-su (present-day Sian), then the capital of the Thang dynasty, I-Hsing developed an escapement that he applied to a water-driven mechanical clock, but the principle was equally applicable to the Western self-contained mechanical clock.

Basically this was simple enough, like so many great inventions, but in practice was a little more complicated. One model consisted of a water wheel with each blade in the form of a pivoted scoop. Water trickled continuously into one scoop at a time and, while this was happening, the scoop – and so the wheel – was held still both by a weight and a weighted arm. Only when the scoop was full was it heavy enough to overcome the weight and trip the arm so that it was free to move downwards. As the arm moved down it tripped a second weighted arm that operated a third arm, which then released the wheel, until the scoop in question had swung away and the next scoop was in position for the whole process to be repeated. This probably sounds most involved, but the artist's drawings on these pages

shows the whole scheme which, like all mechanical linkages, is easier to grasp visually than to explain verbally.

The water mechanism shown here was developed from I-Hsing's basic idea by Su Sung, a Chinese polymath who flourished in the latter decades of the eleventh century AD. He was ordered by the emperor to supervise the reconstruction of an astronomical clock at Khaifêng in Honan – then the capital. This enshrined I-Hsing's escapement, for the wheel was only allowed to rotate (escape) tooth by tooth, and this happened when each scoop was full. Thus the rate the wheel turned was constant, provided, of course, that the water flowed into the scoops at an even rate, and this was achieved, as with Ktesibios's clepsydrae, by using a siphonic filling system.

There is a remote chance that Ktesibios's idea of a constant-flow mechanism may have reached China, but it is just as possible that I-Hsing's siphon was an independent invention. Certainly, though, we have no hint of an escapement in the corpus of tradition passed down in the West, and it was the escapement that made the mechanical clock a practical possibility. Not for 500 years did knowledge of I-Hsing's invention filter through to the West and, when it did arrive, it was hundreds of years before the Chinese accuracy could be matched. Only when a good escapement and a constant regulator – the pendulum – were combined in the seventeenth century, were equal hours universally accepted and the subsequent economic and psychological changes began.

CHAPTER 3

Earth-Explorers and Atomists

INQUISITIVENESS led man to explore every question that his existence posed, from the nature of his soul to the constitution of the ground he stood on. That solid, stable platform he and his forbears had inhabited since the most distant reaches of the past: What was it? What was it like? Was it round or square? Big or small? What supported it? And what lay underneath the ground? All these questions came tumbling out and, at first, the answers to each were legion. The Earth was flat, it was humped, it was square, rectangular, round, it floated on the back of three elephants which stood on the carapace of a tortoise that stood on a giant cobra.

As time passed, it became generally accepted, at least in the Mediterranean area, that the world was some kind of flattish earthy body. To the Egyptians it was the reclining body of the god Qeb and therefore oblong; to the Babylonians round, with a central peak making it somewhat domed or, in some descriptions, very rounded like the upturned spherical bottom of the kufa that they used for general transport on the Euphrates. These shapes were adopted because they were approximate representations of the land mass on which each civilization lived – the Egyptians in a long narrow valley, the Babylonians on a rounder piece of land, with sea, if not at every shore, at least at each cardinal point if one travelled far enough. To the early Greeks the world was a disk floating on water. But gradually these simple, direct interpretations of everyday experience were extended; imagination began to colour the picture, and aesthetic values to play a greater role.

The break with a flat or humped Earth was made by Pythagoras – or should we say the Pythagoreans? – who taught that the Earth was a globe, fixed in the centre of a spherical universe. This was about 500 BC, and from this time onwards the Greeks never wavered in their belief that the Earth was a globe. The medieval idea that it was flat was a retrogression, a loss of the Pythagorean theory which was not revived until would-be circumnavigators like Columbus and Magellan proved by experience that the Earth was round, and the rediscovery of Greek learning made Aristotle's arguments proving the rotundity available to the philosopher. Lost, too, was an astoundingly precise measure of the Earth's size that Eratosthenes made in the third century BC.

Eratosthenes was born at Cyrene (now Shahat in Libya), but after he was thirty most of his time was spent at Alexandria, where he went at the invitation of Ptolemy III. In 235 BC, he was appointed chief librarian when Zenodotos, the first to hold the post, died. Here Eratosthenes remained in office until he was in his eighties. He was primarily a mathematician and geographer, and through a great interest in chronology devised the system of dating events by olympiads. But he was also an authority on Old Greek comedy and is even cited in the scholia (explanatory notes) to Aristophanes's plays. He was obviously a man of wide interests and erudition, but our concern with him centres on his geography, and particularly on

Man's intellectual and scientific development has been accompanied by a gradually increasing knowledge of the world on which he lives. This map of the known world, produced in 1486 but based on that of Ptolemy made some 1,200 years earlier, shows the great expansion in geographical knowledge since the time of the Ancient Greeks. The earliest maps were very parochial, although we now know that explorers made journeys of astonishing length well before the start of the Christian era.

his measurement of the size of the Earth. A word first, though, about his geography in general. It was notable not only for its accurate mapping – Eratosthenes corrected the traditional but erroneous Ionian map that had Delphi at its centre, and showed the Greek world as semicircular in outline, with an outer circular ocean – but also for its then unique mathematical approach to the question of geographical positions.

His measurement of the size of the Earth was simple and elegant. It was based on information that at Syene (modern Aswan) the Sun appeared directly overhead at the summer solstice. At this time, when the Sun reached its northernmost point, a gnomon (shadow-stick) stuck in the ground cast no shadow and the bottom of a well, supposedly dug for this very reason, was completely lit up by a vertical shaft of sunlight. Prompted by these facts, Eratosthenes caused two measurements to be made. One was the elevation, or the angle above the horizon, of the Sun at Alexandria at the time of the solstice, the other the distance between Alexandria and Syene. The altitude of the solstitial Sun at Alexandria coupled with the fact that it was vertical at Syene allowed him to determine the difference in latitude between the two places – that is, the number of degrees Alexandria was north of Syene. Combined with this, the measurement of the overland distance between Alexandria and Syene then gave him enough information to calculate the distance around the whole Earth.

The figure he obtained for the latitude was a little over seven degrees, or one-fiftieth of a circle. The distance between Alexandria and Syene, determined by a bemetatistes, or professional pacer, came out at 5,000 stades. So the Earth's circumference was fifty times this, or 250,000 stades. Actually, although the difference in latitude was pretty correct, Eratosthenes knew that even a professional pacer would not be precisely accurate and, for arithmetical convenience, he took the circumference to be 252,000 stades. This is equivalent to 29,000 miles, and thus a little too large since today's figure is 25,000 miles. He was also in error because Syene was not exactly underneath the Sun at the summer solstice – it was about twenty-three miles too far north – and Alexandria was not due north of Syene, but some 180 miles to the west.

But no matter; Eratosthenes's value for the Earth's circumference was far and away the most accurate figure obtained in the ancient world, and – perhaps more significant from our point of view – it was not bettered until modern times. However, when the Roman empire fell and the Dark Ages arrived, the measurement was forgotten just as the round Earth was forgotten. Bigotry and ignorance prevailed, and only with the recovery of Greek learning was this vital measurement rediscovered. Yet what might the first circumnavigators have given not only to know the kind of world they sailed on, but the distance they had to travel as well?

But if the first circumnavigators were basically unaware of what they were doing, it is not true that there were no major marine explorations before the fifteenth century AD, or that ancient navigators hugged the coasts, afraid to venture into the loneliness of the open sea. We have ample evidence that this was not the case, even though so many early voyages were later forgotten or, at best, became legends exaggerated and distorted in the telling. In the ancient world of the Mediterranean, voyages to the west African coast had been made as early as 500 BC, but the most astonishing sea journeys were those of Pythias almost two centuries later. A younger contemporary of Aristotle and a native of Massilia (Marseille) in Gaul, then a centre for the dissemination of Greek culture, his trip was probably financed by his fellow Massilians. This would be no altruistic gesture on their part, but a sound investment for trade purposes – a motive echoed by British, Spanish and other Europeans many centuries later.

Pythias was known as an explorer, and towards the close of the fourth century BC, with the Carthaginian maritime power just beginning to wane, the Massilians were keen to try to outdo their rivals in trade. They especially wanted to capture the tin and amber trade, and this meant visits to Cornwall and to the Baltic. Pythias set sail sometime between 330 and 300 BC; he passed through the Pillars of Hercules (Straits of Gibraltar), went on to Cadiz, around the Spanish and French coasts – he particularly noted the large size of the Britanny peninsula – and then to St Michael's Mount in Cornwall. From here, the commercial centre of the tin trade, he sailed around Britain, visiting Scotland and Thule (which may well have been Iceland), and on to the southern Baltic.

The voyage took him within the Arctic Circle, and he gave the first recorded description of arctic conditions, and spoke of 'the sleeping place of the Sun'. But in spite of Pythias's audacity and his careful reports, he was generally disbelieved by his contemporaries. His explanations of natural phenomena were also discounted; for instance, his contention that the tides were caused by the Moon was not taken seriously, but whether this was his own idea or whether he was told it by the inhabitants of Scotland – the megalithic circle-builders of Kintyre seem to have known this as early as 1500 BC – we do not know.

Pythias's voyage was not the only distant marine adventure in the world. There is now plenty of evidence from the Kon-Tiki and Ra expeditions of Thor Heyerdahl, and from a study of the stick charts of the Marshall Islands that give details of waves and currents, that the Egyptians and the Polynesians could well have voyaged across the Atlantic and Pacific. It is well established that the Phoenicians had a flourishing maritime trade with west Africa, and the Greek historian Herodotos reported an Egyptian-Phoenician circumnavigation of Africa more than 2,500 years ago. This expedition took place in the time of the pharoah Necho about 600 BC. The Phoenician sailors and their Egyptian sponsors sailed south from the Red Sea and returned through the Gibraltar Straits after three years, having made two landfalls to grow crops. Among other things, the explorers reported that the Sun shone from the north as they rounded the horn of Africa. But a vast corpus of records of other ancient explorations have quite probably been lost.

Perhaps the early civilizations in South America may reveal evidence that Egyptian explorers sailed there in the third millennium

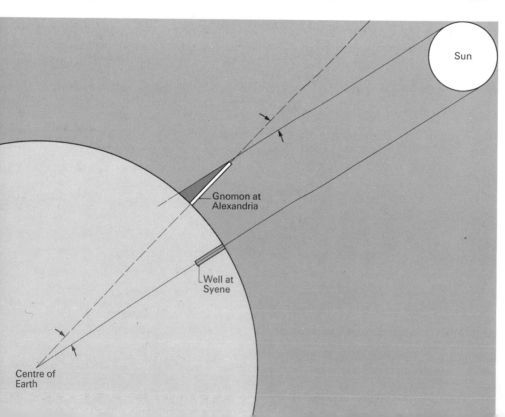

left Eratosthenes measured the circumference of the Earth by finding the distance between Alexandria and Syene and also the difference in angular latitude between the two places. He was able to find the latter by measuring the shadow cast by the noon-day Sun at Alexandria at midsummer. This gave him the angle of the Sun overhead. At this same time, the Sun was vertically overhead at Syene, and cast no shadow at the foot of a well.

Sun

Gnomon at
Alexandria

Well at
Syene

Centre of
Earth

right A map of the possible routes taken by some very early explorers. Pythias probably sailed as far north as the Arctic Circle, while an Egyptian-Phoenician expedition apparently circumnavigated Africa. Also shown are the routes of the two recent Ra expeditions, which showed how ancient sailors could have reached the Americas.

BC, or conversely that there was cross-fertilization of ideas between there and the Mediterranean countries. Possibly Pythias was not the first to penetrate the Arctic – he was certainly far from the first to visit the Baltic to trade in amber – and it may be that the megalithic cul-ture evinced in the stone circles in Britain and Britanny may not have been as isolated as we tend to think. No, it is gradually becoming clear that exploration of the Earth by land as well as by sea was much more extensive in ancient times than was once imagined. The re-discovery of the truth of what Pythias reported, the seaworthiness of even a papyrus boat, of the navigating prowess of the Polynesians, all show that we must be prepared to re-think the whole situation.

So it is clear that ancient man knew that the Earth was round, that he explored it, but, with the odd exception, he thought of it standing still in the centre of the universe. What kept it there, though? By late in the fourth century BC it was the general consensus of opinion that this was its natural place, self-evidently demonstrated by the fact that solid earthy bodies fell towards its centre – which, by definition, was in the middle of all material creation. But what was this centre like? What would one find if one dug down deep towards the central regions of the Earth?

Some indication was given by mines, which were often surpris-ingly deep. For instance, in the Laurion mines, close to Athens, first worked by the Mycenaeans in the second millennium BC and re-opened and extended by the Athenians by 500 BC, shafts were sunk to depths as great as 350 feet. It was soon found that as one descended a mine, the temperature rose – by approximately one degree Celsius

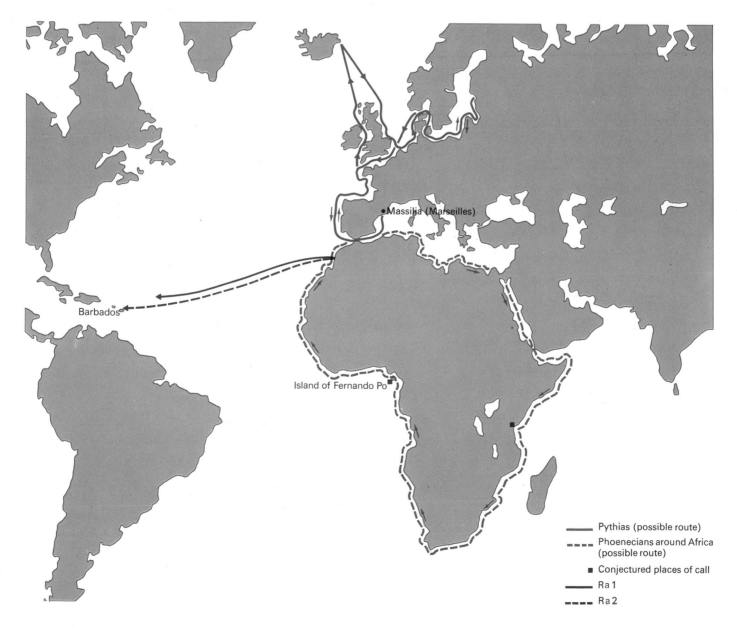

Massilia (Marseilles)

Barbados

Island of Fernando Po

Pythias (possible route)

Phoenecians around Africa (possible route)

Conjectured places of call

Ra 1

Ra 2

(centigrade) for every 90 feet. So at the bottom of one of the deeper Laurion shafts, the miner would not only be working in cramped conditions with rather stale air, but at a temperature some four degrees (or seven degrees Fahrenheit) higher than on the surface. Such a variation in ambient temperatures was enough to be notice able even before there was any method of measuring it. But there were other, and more dramatic demonstrations of subterranean heat.

Hot springs were certainly known after 300 BC, but the most ex citing and terrifying evidence came from volcanoes. There had always been volcanic activity around the Aegean and Mediterranean, and perhaps the best remembered event today is the rebirth of Vesuvius in AD 79. It was during this eruption that Herculaneum and Pompeii were destroyed, and the Elder Pliny was suffocated by its sulphurous gases when he ventured too close out of curiosity. The general idea was that there were great winds locked up inside the Earth and that erupting volcanoes and earthquakes were evidence of these breaking forth, bringing hot rock with them. They were natural phenomena, due to natural forces – not, to the Greeks at least, evidence of divine displeasure. Later this was forgotten, and in the first fifteen centuries of Western Christendom the whole gamut of tempestuous nature became a stage for the display of holy anger. It was not until the late seventeenth century that the rational approach of the Greeks was re applied.

In China somewhat similar views were held about volcanoes and earthquakes, but to the astronomer and mathematician Chang Hêng, who worked in the early decades of the second century AD, must go the credit of having produced the first of all seismographs or, as the Chinese themselves called it, an 'earthquake weathercock'. In essence this consisted of a heavy pendulum which would move only when disturbed by a very slow vibration such as an earthquake tremor would give, and remain unaffected by the quicker everyday type of disturbance – the basis of the seismograph even today. Another feature of Chang Hêng's design was that the instrument not only recorded that there had been a tremor, but also the direction from which it came.

This is not as difficult as it sounds because the direction in which the pendulum moved depended on the direction of origin of the dis turbance, and Chang Hêng's problem was only to find some straightforward indicating system. In the end he arranged that, when the pendulum swung, one of the eight arms attached to it displaced an indicator ball. The pendulum also released a hook which fell and held the arm in position, locking the pendulum, so that the secondary shocks caused by the earthquake would not be recorded. In this way Chang Hêng's seismograph gave an unambiguous result. While sino logists have a good description of the instrument and the way it worked, there is still some doubt about precisely how the pendulum was arranged inside the outer casing, and the details shown on page 51 represent only one possibility.

Like so many scientific instruments up to the nineteenth century AD, Chang Hêng's seismograph hid its utility under a mask of beauty, and the recording of an earthquake was done by causing a ball to drop from the jaws of a dragon into the mouth of a frog. There were eight dragons, each facing a point of the compass, and the arms on the pendulum operated the dragons' jaws. There were, of course, eight frogs sitting agape, one beneath each dragon. This picturesque instrument did not penetrate to the West until our own century, and by then the seismograph had been independently invented some fifty years earlier, following an upsurge of interest in seismology.

Theorizing about the interior of the Earth long ago gave rise to a perennial question that no philosopher could answer with any certainty: How did the Earth come into being, how was it created? One way out of the dilemma was to ascribe it to a divine act – the whim of a god, the caprice of some minor deity, the action of a demiurge. This would put the matter beyond further discussion, at least from the point of view of natural laws; divine creation lay outside the world that was amenable to any scientific analysis of cause and effect. It was an explanation of this kind that did service for Western man up to, and long after, the coming of the Renaissance; but it did not satisfy all the ancient world.

There were some who sought an impersonal solution, who wanted to find a means of creation by natural, not supernatural, forces. Of these the most significant from our point of view was Anaximander, who was born in Miletus on the west coast of Ionia (Turkey) in 610 BC. He thought that while the origin of the entire creation was the Infinite – and this he did regard as divine – the universes in it were created from the Infinite by natural forces, and displayed a collection of 'opposites' that separated out as the universe rotated. The universe was, he believed, shaped like a sphere, and the Earth like a cylinder. But Anaximander's importance lies in the fact that he considered the actual formation of the Earth and the stars to be due to acts of nature, and conceived of them as being generated from a single primeval substance, something subtle and invisible, but a substance nevertheless. This was an immense philosophical stride forward. To find natural forces postulated for the creation of all that the universe contains brings the whole of the material world under one single law. Natural law is universal, natural law knows no physical boundaries, and so the entire gamut of terrestrial phenomena is subject to the same basic laws – a view that the scientist has taken ever since.

Anaximander's concept of a universal underlying substance from which everything arose was not original, for his older contemporary Thales, also from Miletus, thought that water was the basic substance that underlay every other. And even Thales's view was not essentially his own; it seems to have come from Egypt (which he had visited), and was a natural enough idea to hold if one's whole existence depended on an annual inundation. After harvest the land was

A Greek vase painting depicts Odysseus on board a war-galley. The *Odyssey*, attributed to Homer, seems to have been written about the eighth century BC, and tells of the travels and adventures of Odysseus. It contains, perhaps, echoes of real explorations in the Mediterranean and beyond – a mixture of fact and fiction forming legends of dangers and discoveries that were Homer's raw material.

The Roman city of Massilia (Marseilles), as depicted in the *Nuremberg Chronicle* (1493). It was from Massilia that Pythias set out on his great voyage of exploration in the fourth century BC – a voyage that took him well inside the Arctic Circle.

barren, dry, dusty, empty. Only after the Nile waters had risen, flooded the surrounding land, and then begun to subside, did the ground become fertile and green sprigs begin to appear through the surface. Only when water had soaked in and wetted the ground did the plants begin to sprout. Water was the cause – perhaps water was the primeval, essential substance; to say so was certainly logical enough.

As soon as the philosopher looked around him he had to face another problem. Whatever the primary substance might be, there was an abundance of other substances to be explained, a plethora of materials. Did these all come from the primary substance by some strange alchemy, some amalgam of qualities that transformed it into the immense variety of substances seen? Or was there, perhaps, some other explanation? Again, there were strange quirks of behaviour to be accounted for. When some substances were rubbed – amber (Greek *elektron*) was a notable case in point – they attracted other bodies. What was the reason for this? Then there was the lodestone: this drew other pieces of lodestone to it, and also pieces of iron but not of copper – a property that was later to be used as the basis of the magnetic compass. Why was this? Wherein lay the source of this strange power? What made things hot? Why did fire burn? What was its nature? How could all these qualities and varieties found in the natural world be reconciled? Was there one hypothesis that would account for them all? Or was it necessary to have a series of explanations for so many different things?

We are not sure what the Babylonians thought, although it is hardly likely that they had no opinions. But the cuneiform tables that we have say nothing about these matters, and their detailed views are lost. The same applies to any theories the Egyptians may have had. Our knowledge starts with the Greeks and, to begin with, there seem to have been many answers to every question. The philosopher Heraklitos in the late sixth and early fifth centuries BC extolled fire as the primary matter; all other substances were derived from it. Anaximenes, on the other hand, believed air to be the basic substance of the universe. Gradually, however, a certain attitude emerged; there was an attempt to marry qualities and varieties together. There was also, in the hands of Aristotle, a differentiation between the quality of heat and the elementary substance 'fire'. Indeed Aristotle's writings on the whole question of natural materials formed the basis of what was later to become alchemy and then, later still, to be transformed into the science of chemistry. And they were ingenious enough, explaining as they did the majority of chemical phenomena, although they omitted any mention of either the electrical attraction of amber or the attractive power of lodestone.

According to Aristotle there were four primary elements, and four basic qualities; from these the whole amazing array of natural materials could be constructed. The four elements were earth, air, fire and water; the qualities hot, cold, wet and dry. These elements were

Recent research has shown just how seaworthy even apparently primitive craft could be. **above** An Egyptian tomb painting shows a boat made by binding together bundles of papyrus stems. Such boats were common in Ancient Egypt, but until Thor Heyerdahl sailed *Ra II* across the Atlantic, most people believed that they could be used only for river or coastal journeys.

below A Maori war canoe, drawn by a member of Captain Cook's expedition to New Zealand in 1769. The Maoris are Polynesians, who spread across much of the Pacific before the era of European exploration. They made enormous voyages in canoes like this, and current research shows that they were aided by an efficient method of astronomical navigation.

not the ordinary everyday earth, air, fire and water with which we are familiar and which he considered to be gross combinations of the elements. They were the subtle elements that lay behind them. Here we have an echo of Plato's teaching that the material world is but a shadow of an essential underlying reality which we can never touch.

For Aristotle each element had its natural place in the universe. The centre of the Earth was the natural place for the element earth, the air around the Earth the natural place for air, the sphere above the air (but not as far away as the Moon) the natural place for fire – this is why flames always burn upwards – and all around the Earth lay the natural place of water. An earthy substance fell to the ground because it was seeking its natural place, a burning log would have flames that burned upwards and a residue that fell downwards because it was composed of both earthy and fiery elements. Sunlight produced the quality dryness and caused dry 'exhalations' from the substances both earthy and fiery, and if these exhalations could not escape, then rocks were formed. Sunlight could also cause wet exhalations if substances were combinations of the elements air and water; if these exhalations were not free to escape they formed metals – after all, were not metals cold and clammy to the touch? From the sphere of the Moon outwards different rules applied, for all celestial bodies were composed of a fifth essence, eternal and incapable of blemish or decay.

Aristotle's scheme was not only ingenious; it collected under one umbrella a whole host of facts of everyday experience. Not surprisingly it became widely accepted and, with his belief in possible transmutation of the four elements, led directly to the pseudo-science of alchemy. Nor was this doctrine of the four elements and the four qualities lost. It was developed and extended, via Alexandria and the alchemist Zosimus, to the Christian and Muslim worlds. It was not forgotten, and did not have to be rediscovered; rather it had to be exorcised from Western thought. Yet it has been necessary to describe it simply because it was so pertinacious a view, so deeply woven into the fabric of chemical thought, that when the atomic theory was pro-

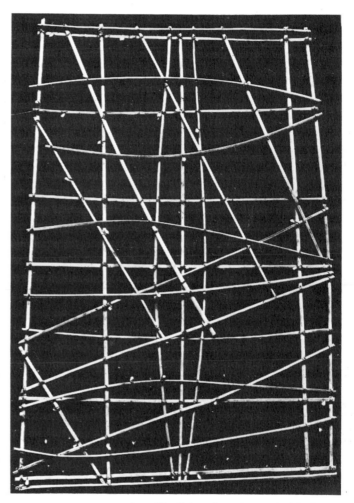

above The Polynesians also used stick charts such as this to guide them on the vast, almost empty Pacific Ocean. This is a chart of water movements, not of land. The navigator would observe the relationship between the main waves driven by the trade winds and the secondary waves reflected from an island. In this way, they could find their destination.

posed by John Dalton in the early years of the nineteenth century AD, much prejudice had to be overcome before it could be critically examined in an unbiased way. Yet Dalton's was not the first atomic theory. The idea that matter is composed, not of four mystical elements, but of myriads of tiny particles, has its origin in Ancient Greece, in the fifth century BC, over 2,000 years before Dalton was born.

The first person known to speak of atoms was Pythagoras, who thought of them as tiny cubical shapes, but Greek historians of later centuries claimed that an atomic theory was much older, and some of them said it came from Babylonia. At all events, we are concerned with two Greeks of the late fifth century BC, Leukippos and Demokritos. Leukippos was another native of Miletus, and he laid down the main outlines of a sound atomic theory, outlines that were developed by his contemporary Demokritos. The latter was born in Abdera on the north coast of the Aegean.

Since the theory is inextricably linked with both men, there is no need here to try to sort out what were Leukippos's original proposals, and what were Demokritos's specific contributions; it is the concept as a whole that matters. According to them, the universe is composed of two parts – atoms and void. A void, a nothingness, a vacuum, was something that many philosophers disliked, and whose existence Aristotle was later to deny. Yet if one was to have a particle theory it was, if not a necessity, at least a highly desirable condition; otherwise one would have to suppose that the atoms moved about in something that could not really be called a substance and was merely a philosopher's fiction. Leukippos and Demokritos rejected any compromise. For them the universe and our Earth in it were composed of atoms and void, and nothing else whatsoever.

The atoms were small, most, if not all of them, invisible, infinite in number and of a multitude of different shapes. They differed in size. But although they were solid and real enough, they did not display any of the ordinary qualities to be found in everyday material bodies. And they were indivisible, unable to be broken down further – indeed, that is the essential meaning of the Greek word *atmos* (uncut). They had no holes, no dents, but were one solid piece of essential substance, although it seems that in general they were thought to be hook shaped so that they could lock together. All atoms were in continual motion (a pre-echo of the modern kinetic theory of matter?). And when a group of atoms became isolated it

left Even in antiquity, mines were dug to considerable depths, as is shown by this illustration from Agricola's *De Re Metallica*, published in the middle of the sixteenth century AD. These mines were similar to the Laurion mines, near Athens. So early man had a direct demonstration that the centre of the Earth is hot, for the temperature was noticeably higher deep underground.

above An even more dramatic demonstration of subterranean heat came from volcanoes, such as Vesuvius, whose eruption in AD 79 destroyed Pompeii and Herculaneum. The Greeks took a rational, scientific view of these as being due to natural forces – in stark contrast to the later superstition that blamed volcanoes and earthquakes on divine displeasure.

whirled round, and in this way similar atoms collected together.

In this way our particular universe was formed, and the substances in it. And interestingly enough, considering our present twentieth-century ideas of the significance of the arrangement of atoms in molecules, whereby two different arrangements of the same atoms give substances that are chemically identical, but quite different in their reactions with other materials, so these Greek atomists claimed that their atoms could be arranged in various ways to give various substances. The space between atoms was subject to infinite variety, and the qualities observed in different materials depended both on the atoms, their arrangement, and the distances between them. Demokritos himself even went so far as to apply the atomic theory to biology, but, most unfortunately, nothing he wrote on this has come down to us, either directly or in the way of comments. We are, therefore, in the dark about what he thought of the atomic nature of living material, and how he explained its interactions and the behaviour of individual animals and plants.

Unlike the atomic scientist of today, neither Leukippos nor Demokritos confined their atomic theory to the material world. The soul was made of atoms, too – spherically-shaped atoms that moved about at high speed, but atoms all the same. In essence, then, the soul was no different from the body; when one died the atoms of the body were dispersed, but not destroyed, and so it was with the soul. Yet this meant that there was no existence of an individual soul after death, only of its separate atoms. And there was another extra-scientific aspect of the Greek atomic theory. This was concerned with the movements of the atoms themselves: not just their ceaseless motion, but the movements that occurred during chemical reactions, when one cut something up, whenever, in fact, objects underwent change. Such motion, they said, was due to the 'inner necessity' of the atomic substances, the force of the physical world, and this led directly to a

below Probably the earliest printed illustration of an earthquake : the fall of Babylon from the Nuremberg Chronicle.
right The first seismograph. Invented in China early in the second century AD by Chang Hêng, this had a pendulum heavy enough to respond only to the slow vibrations of earthquakes. The pendulum swung in the direction from which the tremor came; this made the appropriate dragon's mouth open and release a ball which dropped into the mouth of the frog below. A catch on the dragon release mechanism locked the pendulum so that it could not swing back and release another ball. These drawings are based on a reconstruction by Wang Chen-To discussed by Dr Joseph Needham.

Dragon's head

Pivot

Lever

Clutch

Crank

Arm

Ball

Pendulum

Frog

Falling ball

materialistic determinism. The whole of life, spiritual as well as physical, was regulated by atomic behaviour.

The atomic theory as Leukippos and Demokritos conceived it was a bold, far-reaching hypothesis that could have affected man's whole outlook. It was staggering in its simplicity and in its power to explain the natural and supernatural worlds in one all-embracing scheme. It was one of the greatest imaginative conceptual hypotheses of the ancient world, but it did not commend itself to later generations. In a century and a half it was eclipsed by the Aristotelian scheme, and was relegated to the class of an interesting but unacceptable theory. And even 200 years later, in the first century BC, when it was revived by the Roman poet and philosopher Lucretius, who enshrined its doctrines in his magnificent narrative poem *De Rerum Natura (On the Nature of Things)*, it was rejected. Even though Lucretius's poem was praised, its contents had no effect on the scientific world. The West received Aristotle's alternative doctrine.

left From the *Nuremberg Chronicle*, a representation of Aristotle, who was to become perhaps the most influential of all the Greek philosophers. His theory of the four primary elements — earth, air, fire and water — was a brilliant synthesis in its time and was firmly materialistic. But it misled medieval scholars and had to be painfully exorcised from Western thought. **left above** A sixteenth-century illustration of the creation (from the *Margarita Philosophica* of Gregor Reisch, published in 1508) shows the influence of Aristotle's ideas: God is creating Eve out of Adam's rib. This takes place on a (flat) Earth, which is surrounded by water. Above the sphere of air is shown the sphere of fire.

left An apocryphal portrait of Anaximander, also from the *Nuremberg Chronicle*. This Greek scholar, who flourished in the sixth century BC, was among the most important natural philosophers of the ancient world. He was one of the first to insist that the creation of the Earth was a natural, physical process, and not a divine act.

CHAPTER 4

Mathematics and Computers

TRADITION has it that on the door of Plato's Academy in Athens was the inscription, 'Nobody should enter who is not a mathematician.' The story has the ring of truth. Plato extolled the virtues of a training in pure mathematics, and would have liked it taught to every embryonic statesman; God, he said, was always geometrizing. But practical mathematics was a different matter. Plato despised such everyday things as the business of counting and the application of geometry to measuring areas of land; these were degrading pursuits, and the philosopher should concern himself with pure numbers, with contemplating their relationships, and so directing his soul to higher thoughts. Yet, although numbers attached to visible and tangible bodies were anathema to him and smacked of the market place (where, indeed, they were a necessity), it was in trade and simple surveying that mathematics had its roots. Whatever philosophical exercises numbers and shapes might provide, they arose because of man's need to count, to number, to separate and divide. They were the commerce of the administrator, the tax gatherer, the trader, the builder and, in short, the whole body of practical men without whom the philosopher would have starved, unhoused, unclothed and un-cared for.

Who first began to count we do not know. The very beginnings of numeration are lost in the mists of prehistory, but recent research has made it very evident that the Sumerians had already developed the art to a high level as early as the third millennium BC, and the lead they gave was taken up by their successors. Certainly, much of what they did was done for purely practical reasons, but a small amount seems to have been done for the love of the subject, and had little if any immediate practical application. But how are we to describe it? Most people fight shy of a mathematical equation, and their normal intelligence seems to fly out of the window as soon as a series of numbers appears on a page. This may be due to poor teaching; it may also be due to having no head for figures. But since we shall be concerned mainly with numbers and shapes, the problem should not be too difficult: Everyone can count and no one will fail to recognize the difference between a square and a circle.

It is one thing to count out loud, but quite another to have a method of writing down the results. Small numbers – one, two, three or four – present no difficulties; a number of dots or dashes will do. But as soon as we go on the number of signs becomes so great that rather than recognize a number at a glance, one has to start counting the signs all over again. The Babylonians and Sumerians realized this basic fact well enough. In their cuneiform writing they used a wedge-shaped dash for one, two dashes for two, three for three, and then began to write the dashes again, so that, for example, the number nine appeared as three rows of three dashes each. This meant that one could see at a glance what the number was. Ten was different; ten consisted of two wedge-shaped dashes, one on top of the other, and so was distinctive. Twenty was two tens, thirty was three tens, and so on up to fifty; after that it was too much and the eye became confused. But they did need to count much higher than fifty. How could they do this and still have instantly recognizable numbers?

The Sumerian answer was simple but brilliant, directly to the point and, at the same time, of profound significance for all their arithmetic and for all their subsequent mathematical work. They adopted place value notation, using powers of sixty. This sounds dauntingly complicated, but it is merely what we do now, only we use powers of ten instead of sixty. The Babylonians shifted their numbers one position to the left at 60, then another step to the left at 60×60 or 3,600, and again at $60 \times 60 \times 60$ (21,600), and so on. But since this sexagesimal method is unfamiliar, it may be a help in grasping the Babylonian system if we use our own decimal method as a working example. We have the single symbols 1, 2, 3 and so on up to 9. When we get to ten we write the digit 1 again, but shifting it one place to the left. To show that it *has* shifted to the left, we write in a zero after it, so we get 10. Then we go on again, just as the Babylonians did; eleven is 11, twelve is 12, and so on. And when we arrive at 10×10 we again shift to the left, so that one hundred is 1 moved two places to the left, or 100. To show position, the Babylonians also used a zero, only theirs was not a circle; that would have been too difficult in cuneiform script. Instead, they had a sign rather like Σ.

This place value notation was of cardinal significance because it enabled a mathematician to handle numbers, to prepare tables of numbers, and to see at a glance the relationships between one number and another. The Egyptians had no such place system for writing numbers, no sign for zero. What is most remarkable, neither did the Greeks or the Romans. We can appreciate how difficult and cumbersome this made things if we use Roman numerals and try subtracting CXCII from CCCCVI; it is nowhere near as easy as finding the answer using the so-called Arabic numerals, where we write 406—192, to give us the answer 214 almost straight away. And it does not matter what base one chooses – ten, sixty, or some other number – the principle is the same and facility comes with practice. If we were brought up to count in sixties instead of tens, we should find it just as easy; after all, we do it quite happily when it comes to time, when we say that sixty seconds make one minute and sixty minutes one hour.

In this sense we are more heirs of the Sumerians than the Greeks

The annual flooding of the Nile valley and the consequent need to survey the farmland that was re-exposed after the floods receded led to the development of a great deal of practical mathematics in Ancient Egypt. This was in spite of the lack of a convenient number system with place values, such as we use today and such as existed in Babylonia.

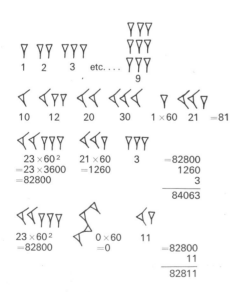

above This tablet is an Old Babylonian mathematical text dating from about 1600 BC, and shows the use of cuneiform numerals. **below** These were written to the base sixty rather than the base ten that we use today. For more details, see the text of this chapter.

and Romans ever were. Indeed this brilliant Sumerian/Babylonian mathematical 'invention' was lost as far as the West was concerned and was only recovered when, in the Renaissance, we learned of the method from the Muslim world. This is why the numbers we now write came to be called Arabic numerals. In fact, to be fair, the Muslims learned them from the Indians who, it seems, had developed them – as well as a position system and a zero – possibly as early as 200 BC. This was much later than the Babylonians, and scholars are divided over whether the Indian system was indigenous or imported from outside – perhaps even from Babylonia.

The Babylonians also had the beginnings of what later came to be called algebra, an Arabic word because, again, the West learned it from the Muslims. Here they did not write in the way we do now, using x and other letters from the alphabet in place of numbers, but they did deal with equations, and in a most uninhibited way. To see how freely they developed their ideas, consider what we can do with x and what the mathematician delights in calling the powers of x. Suppose x represents a length – the length of a piece of string, if you like, or the side of a clay tablet, or anything else that is straight. Now suppose we want to measure the area of a square tablet; then, since a square has sides all of equal length, its area is x times x, or x^2, or the second power of x. For example, if its sides were 3 inches in length, then its area would be 3×3, or 3^2, which gives 9; 9 square inches is the area.

But the use of x – or the Babylonian equivalent – does not restrict one to a particular number and, as will now become evident, this was to have important consequences. For the Babylonians dealt with x and x^2 and x^3 just as if they were mathematical entities, and so were able to handle equations involving x and x^2 (quadratic equations), and equations with x, x^2 and x^3 (cubic equations). This allowed them to deal mathematically with all kinds of problems in ways that need not detain us, but ways that were not available in the medieval West. Here it was thought impossible to handle equations with x^2 and x, since x^2 was an area and x a length. By divorcing themselves from purely practical mathematics and dealing with the subject on an abstract plane, the Babylonians had no such curbs on their imagination. They even used x^4 when convenient, although they could not picture what, if anything, it signified.

And the Babylonians did not stop there; they used the concept of negative numbers, another example of their being quite untrammelled by any need to form a pictorial representation in their minds of what they were doing mathematically. It is little wonder, then, that when it came to applying their mathematics, they were early able to devise something like their zig-zag numbers to describe planetary positions (see Chapter 1) without having to tie themselves to any particular theory of planetary motion. Unhappily, the Greeks did not follow in this algebraic tradition; theirs was a geometrical genius and although they, too, made great strides, their lack of algebra held up

above A sixteenth-century engraving shows the two main calculating methods. The female figure represents Arithmetic. On her left, Pythagoras uses a counting board – a kind of abacus – on which he has formed the numbers 1241 and 82. On Arithmetic's right, Boethius uses Arabic (in fact, Indian) numerals. Medieval scholars wrongly thought the two men invented these methods.

left The Great Pyramid of Cheops, built more than 4,500 years ago, was planned with geometrical precision. But the Egyptians had no theoretical geometry, and had to devise rule-of-thumb methods. For example, they could not calculate pi, the ratio between a circle's circumference and its diameter. So they measured distances that would normally involve calculating pi by rolling a large stone disc cut to a certain diameter.

right A fragment of an Arabic geometrical text. The Arabs developed algebra, but used the geometry of the Greeks. In fact, the Christian West re-learnt its Greek geometry via the Arabs.

below The Greeks developed many geometrical techniques, and this medieval illustration shows some of their practical applications, including making astronomical observations, measuring the Earth, surveying and building. But the West lost its full knowledge of Greek geometry in the Dark Ages, and had to regain it via Arabic scholars in southern Spain.

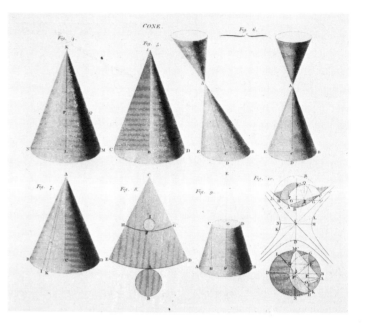

one whole side of the development of pure mathematics. It was this penchant for geometry that the West inherited and that retained its hold on men's minds well into the seventeenth century AD. This was in spite of the introduction of algebra from the Arabs in the twelfth century.

As well as their purely theoretical exercises, the Babylonians naturally applied their mathematics to everyday life, and once again they produced useful and novel things. They devised tables that allowed the practical man to work out common-or-garden problems without needing a deep theoretical background, and they had a sound method for directly calculating compound interest on loans, never the easiest thing to do. Meanwhile, across to their west, in Egypt, the practical side of arithmetic and geometry was the sole mathematical development; there was no attempt there to produce a basic logical discipline as the Babylonians did. All the same, the Egyptians worked out some excellent practical mathematical tricks, especially when it came to handling solid geometry – as they were forced to do in building monuments like the pyramids.

In the construction of a pyramid, the slope of the sides depends on the overall height and the width of the base. But no pyramid was built with smoothly sloping sides to begin with; this was not practicable. Instead, the whole structure had to be made of stone blocks which gave a stepped appearance; only when this had been done could the gaps – the steps – be filled in and plastered over. When we look at the heights and bases of pyramids that were built, and consider the fact that a stepped stone foundation had first been constructed, it becomes clear that there was just one way to have laid out the site correctly before work began. This was to measure off distances in a way that involved the relationship between the diameter of a circle and its circumference, a relationship that involves a quantity we call pi (π). This is not as mysterious as it sounds, for if a circle has a diameter of three feet, say, then its circumference is three times π; if its diameter is four feet, then the circumference is four π. What does raise a problem is π itself, because it is not a simple number, but what mathematicians delight in calling a transcendental number: It has no exact value but is equal to $3 \cdot 14159 \ldots$ In other words, it is not a whole number but an endless series of non-recurring decimals, and so transcends the ability of our number system to express it precisely.

A question that has always been asked is that, since the Egyptians needed to know this before they could build their pyramids, did they possess some secret knowledge that allowed them to calculate it? The answer seems to be emphatically no, and it has been suggested recently that there is a simple solution to the question of how they used π in laying out the base before they started building: They used a stone disk to do the marking. They placed it on its edge next to a peg stuck in the ground, and then rolled it along so that the stone revolved once. There they put in a second peg which was thus exactly π times the stone's diameter from the first peg. And they could do this as

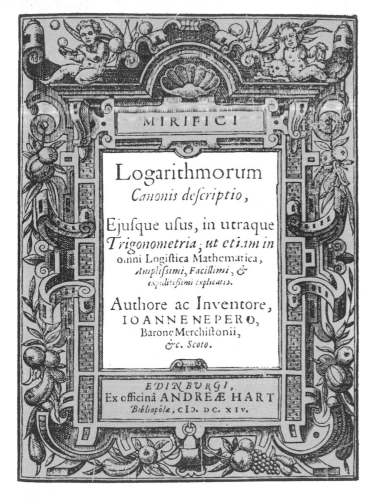

bottom A sixteenth-century illustration of Archimedes in his bath, where he is said to have solved the problem of King Heiron's crown. left The title page of the book in which John Napier explained the use of logarithms as an aid to computation – a method that had been foreshadowed in the fourth century BC by the Greek mathematician Aristoxenos.

many times as they wanted, even though they did not know the exact value of π. By choosing, or cutting, the diameter of the stone so that it was a definite fraction of the pyramid's height, they could obtain any ratio between height and base that they wanted – a direct and practical way out of an awkward piece of mathematics that shows Egyptian pragmatism at its most ingenious.

It is to the Greeks, however, that we must turn for some of the most advanced mathematics of antiquity. Much of what they did came down directly to the West, but by no means all. Their methods of deduction and proof were handed down in an unbroken line, enshrined in Greek geometry as expounded by Euclid, perhaps the most famous of all mathematicians. But what we are concerned with here is the material that disappeared for a while, the material that was for a time lost to our civilization and only arrived in the twelfth century AD or later. Amongst this was some of the work of Archimedes.

Archimedes was born in Syracuse sometime about 287 BC, and it was here he lived and worked until 212 BC. Then the Romans under Marcellus besieged the city and, it is reported, were hampered by Archimedes's resourcefulness in inventing machines for its defence, notably catapults and concave mirrors for reflecting and concentrating the Sun's rays to set the enemy ships on fire. It was during the capture of Syracuse that Archimedes was killed, so it is said, while he was contemplating some geometrical figures he had drawn on the ground. Surprised by a Roman soldier, Archimedes shouted to him to keep off the drawings, and was promptly despatched; a plausible if unconfirmed story. As far as Archimedes himself is concerned, the philosopher and biographer Plutarch claims that although Archimedes's inventions brought him an immense reputation, he regarded the whole business of mechanics as sordid, and no better than any art directed to use and profit; his real interest lay in subjects untainted by any practical utility.

Whatever we may think of it, this was a typically Greek attitude, yet ironically it is for his inventions, mechanics and hydrostatics that Archimedes is remembered. The Archimedean screw for raising water was ubiquitous in Mediterranean countries, and is still to be found where modern technology has had little impact. And his solution of the problem of King Hieron's golden crown, which had a utilitarian end, has become inseparably linked with Archimedes's name. But perhaps he would not have minded so much about the crown, for it helped him lay the foundations of the theoretical – or 'philosophical' as he would have called it – study of hydrostatics, and it is still worth recounting as an example of the way his mind worked. Hieron II, king of Syracuse, was a kinsman and friend of Archimedes, and he suspected that a gold crown he had commissioned was adulterated with silver. Was this true? And if it were, how great was the amount of silver? Archimedes is said to have solved the problem while in a bath tub, when he suddenly became aware of the

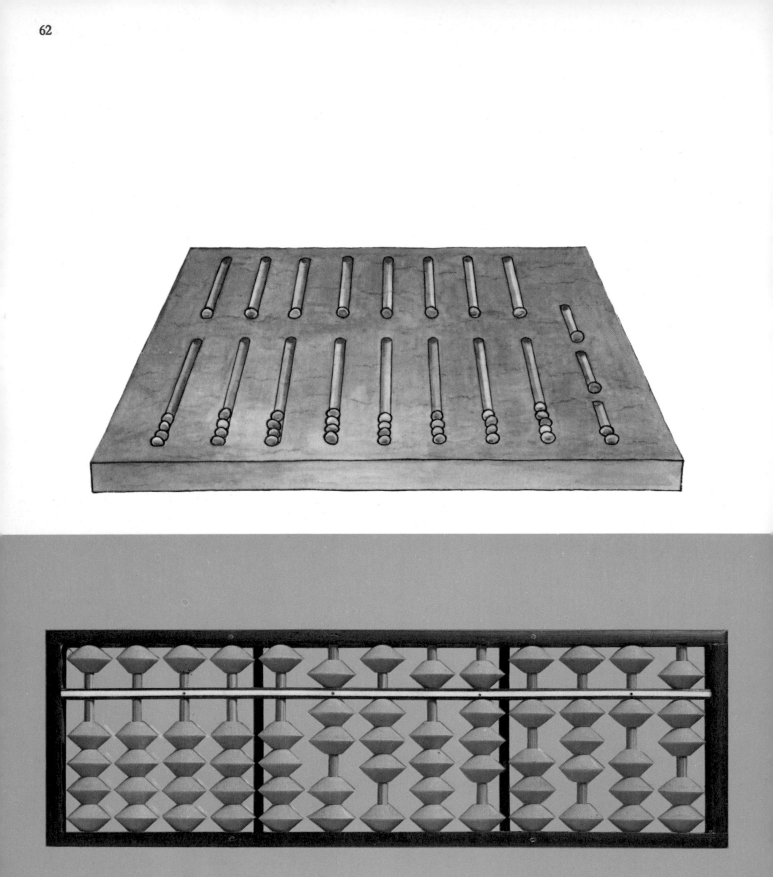

top A Roman abacus generally consisted of a board with grooves cut in it, in which small pebbles — known as *calculi* — were placed. There were grooves for ones, fives, tens, fifties, hundreds and so on. **bottom** The present-day Japanese abacus, or *soroban*, is a much later invention, but is still widely used in the East. It is slightly different from the Chinese abacus, which preceded it. The number shown here is 73,581,317.

right An artist's reconstruction of a Greek astronomical computing device discovered in 1900 in a wreck off the island of Antikythera. It contained a complex system of gears, and seems to have indicated the positions of the Sun and Moon, and perhaps the planets also. This drawing is based on the studies of Professor Derek de Solla Price.

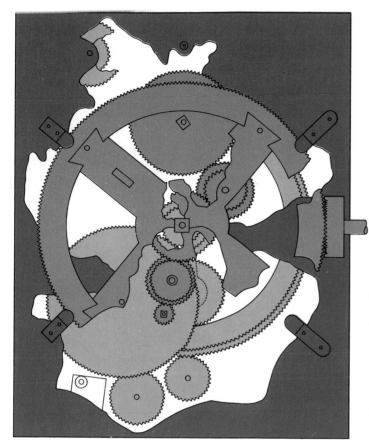

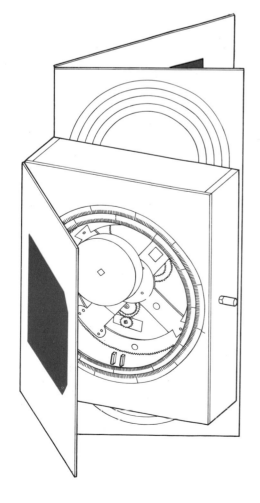

buoyancy of his own body, and he leaped out, shouting 'Heureka, heureka' ('I have found it').

At all events, Archimedes discovered not just the answer to King Hieron's question, but the basic principle of hydrostatic weighing: A body is weighed on a balance in the normal way, and it is then immersed in water and weighed again. The second time it will weigh less because it is buoyed up by the amount of water it has displaced. The difference gives a measure of the weight of the body compared with the weight of its own volume of water. A gold crown would show a greater difference than a silver or an amalgam one, because gold is heavier, volume for volume, than silver. So Archimedes found that he could carry out his analysis without melting the crown down.

His work in mechanics was, again, not only practical; he drew out theoretical principles and so laid the foundations for later generations. He discussed in detail the reasons behind the power of a lever, as epitomised in his legendary boast to King Hieron, 'Give me a point of support and I shall move the world.' But not all Archimedes's work on mechanics came through to the West; his treatise on the basic methods of treating mechanical problems geometrically – which was, incidentally, dedicated to Eratosthenes – did reach Byzantium (Istanbul) in the form of a parchment copy, but it was later erased and written over when the cost of parchment rose prohibitively. It only came to light in 1906 when it was discovered underneath the description of a religious ritual. Again, Archimedes's treatise on the regular heptagon (a seven-sided figure with each side of equal length), written originally in Greek, has been lost, and even an Arabic translation was unknown until one was discovered in Cairo and published in 1926.

By and large, Archimedes's mathematical texts were not popular, and there seems to have been no concerted effort to preserve manuscripts of them as there was with the works of Aristotle or even Euclid. And this is the greater pity because Archimedes came close to some modern techniques. For instance, he spent much time and thought on the very difficult problem of finding the areas enclosed by all kinds of curved figures on a flat surface, and the still harder one of computing the volumes of spheres, parts of spheres and other much more complex solids. His way of tackling these problems – for he treated them separately – was to divide each figure or solid into an immense number of minute pieces, find the area or volume of each of these, and then add them together: a technique that was as revolutionary as it was basically simple. But this is basically the technique that mathematicians adopt for solving the same problem today, only they use the more convenient methods of algebra instead of mere counting, and call the process integration. Unfortunately, this product from a man who, historians are now clear, was nothing less than a mathematical genius, was virtually unknown in the West, and the significance of his achievement was unrecognized even to a geometrically-slanted post-Alexandrian world. The advantages of this method of integration by the summation of small areas was not allowed to help mathematics develop as it might, and for some eight centuries it remained ossified.

Another Greek mathematical method which might also have exerted a strong generative influence, but which did not, was devised by Archimedes's younger contemporary Apollonios. He was born in Perge (in south-east Turkey), worked in Alexandria and later in Pergamum (in western Turkey), and wrote almost as much as Archimedes. Unlike Archimedes, however, Apollonios is famous primarily for one book, the *Conics*, which (as its name implies) deals with cones and all the curves that can be generated by slicing sections through them. These curves range from the circle and ellipse to the parabola and hyperbola. It was not the first book on the subject, but it was far and away the most comprehensive, and was divided into eight sections. Not all of these have survived. The first four sections have come down in the original Greek, but the next three are available only in Arabic translations. The final section has never been recovered in any form, although in the eighteenth century the astronomer Edmond Halley reconstructed a version of it, based on the previous sections and on comments by the third-century AD mathematician Pappos, who had possessed a full Greek text.

What was the importance of the *Conics*? Surprisingly enough, its full significance was not realized for a very long time after the third century AD, and none of the pregnant ideas in it were developed by mathematicians until the seventeenth century. In particular, book five concerned itself with problems that have since become called 'maxima and minima', discussions in this case of the longest and shortest distances from a given point to different parts of a curve. Mathematically this was highly interesting, and since it is a type of problem that is ideally suited to the technique of the calculus developed by Newton and Leibniz in the seventeenth century, Apollonios has sometimes been called their forerunner, or even the original inventor. But this is an exaggeration; Apollonios did not invent the calculus in the third century BC. But if his geometrical work on maxima and minima had been fully appreciated, and if it had been followed up, then probably at least some geometrical form of calculus would have been invented instead of there being a gap of almost two millennia.

Although the Greek mathematical genius showed itself mainly in geometry, there was some interest in numbers from a philosophical point of view. Pythagoras immediately comes to mind in this context, because he linked numbers with shapes, thereby working out what we should now call series relationships, and considered numbers as an important aspect of the harmony of the universe. As an accompaniment to his divine harmony, he examined the relationship between the lengths of vibrating strings and the different musical notes they produced; this was a study that was to bear fruit in the hands of Aristoxenos three centuries later.

Aristoxenos was born in Tarentum sometime between 375 and 360 BC, and was taught by one of Pythagoras's followers, as well as by Aristotle. He is mainly known to us for his *Elements of Harmony*, which contains some original work on musical intervals. This is of considerable interest in the light of twentieth-century music, because

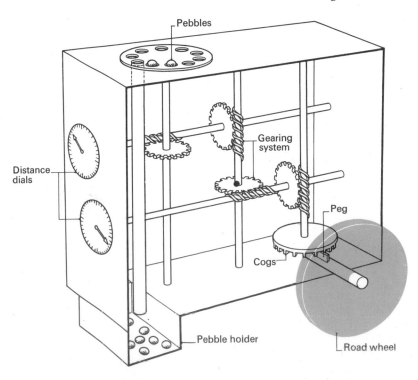

This Roman distance-measuring device – a hodometer or 'taximeter' – was described by Vitruvius. A series of gears translated the movement of the road wheels into the slow rotation of a disk which allowed pebbles to fall into a container. The number of pebbles indicated the distance travelled. This could be checked with dials on some models. Are such devices as this evidence of an ancient advanced technological sub-culture about which we know little?

Pebbles

Gearing system

Distance dials

Peg

Cogs

Pebble holder

Road wheel

he investigated not only the intervals of a tone and a semitone, but even got down to microtones. Mathematically he had to calculate all these intervals and their relationships, and the significant thing is that he did so by a technique of adding his results that was equivalent to our modern method of logarithms. In other words Aristoxenos was able to save himself a great deal of tedious multiplication by a process of adding. Logarithms were devised in the West in the early years of the seventeenth century, yet if Aristoxenos's book had been the musical text of Western Christendom, they might have come much earlier. However, the West learned its music by way of Boethius's sixth-century Latin text based directly on Pythagoras; Aristoxenos's ideas were only taken up by the Eastern Christians centred on Byzantium, and so, in practice, were virtually lost to Western culture.

With the Greek love of geometry and Plato's disparaging views of applied science, one might think that the Greeks would never have been interested in mechanical calculating devices. But a careful sifting of the evidence, scattered and incomplete though it is, makes it seem that this old view may not be right after all. Of course scholars have always known that the abacus was widely used to speed up calculations; in the form of counters sliding along wires it is still in everyday use in many eastern countries, although in the ancient world it was sometimes merely a board covered with dry soil that was marked with changing dots as computation proceeded, or a table with lines on which counters were moved. However, in 1900 some important new evidence arrived. Greek sponge divers, anchoring off the island of Antikythera, south of the Peloponnese, during a storm, saw an old wreck. They went to investigate, and what they found was enough to bring archaeologists to the scene.

After a careful examination, it was clear that this must have been a commercial vessel, sunk sometime about 65 BC while on its way to Rome from either Rhodes or Cos. The cargo was mainly pieces of sculpture, but among the items recovered was something quite different, a device of bronze plates with complicated gears and engraved scales. Subsequent study has shown that it was designed to display the positions of the Sun and Moon, and possibly the planets as well. This might not have been particularly remarkable if it had been a model of the celestial bodies operated by a gear train, a kind of mechanical orrery; but it was not. It was an instrument that gave the positions of celestial bodies in figures – there were pointers that moved over dials to indicate the results of its internal calculations. In fact, it was a kind of mechanical version for displaying what amounted to the Babylonian zig-zag position numbers. In short, this was a mechanical computer, and a complex one at that. Internal evidence also shows that it was a contemporary machine definitely made for everyday use, and not a treasure from some bygone age. We are forced to the conclusion that it pays tribute to a tradition of highly advanced technology in Greece, a technology never mentioned by the philosopher, whose interests were theoretical and not practical, but one that existed nonetheless, even though it was lost.

Perhaps the whole concept of a parallel technological culture is not so unlikely when we consider that the Romans, like the Egyptians, were immensely practical, and made use of various mechanical devices. They had powerful cranes, efficient pumps and a great collection of military machinery, and from what the first century architect and author Marcus Vitruvius said, a flair for producing useful devices. For instance, distances were often measured using a 'taximeter' or hodometer. This had a wheel that ran on the ground, and a series of internal gears that finally ended up turning a perforated disk which released pebbles into a container at regular distances. These could be counted at the end of a journey to assess the distance, while the taximeter seems frequently to have been built with dials and pointers to indicate the answer. But if there was a tradition of an advanced technology in the ancient world, it did not penetrate into Western Christendom.

CHAPTER 5

Studying Life

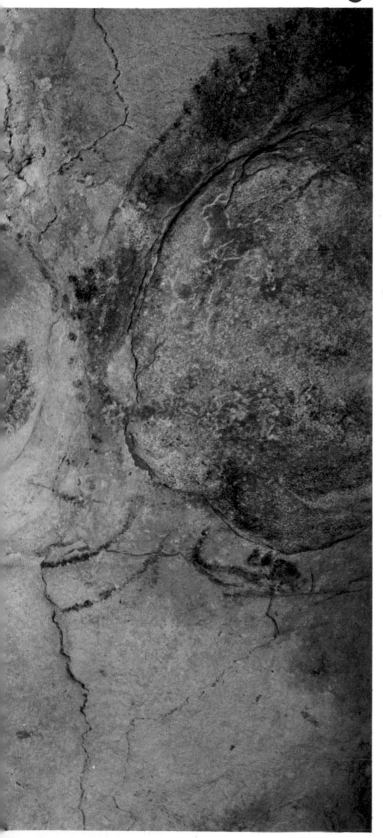

THE range and variety of living things make up an endless and confusing collection of seemingly illogical forms. Sometimes there is a superficial similarity, but when one probes deeper two organisms prove to be poles apart. Sometimes there are no obvious similarities to be seen in what turn out to be closely related species. So the philosophers of the earliest civilizations were brought face to face with a problem that was to trouble their successors millennia later: how to classify and organize the immense amount of material around them, and how to explain the different vital processes they observed. And to make their task more difficult they were often thrown off course, either because some animals and plants were sacred and had to be treated separately from the rest of the natural world, or because they were too uncritical of the fabulous tales brought back by travellers – they just did not have the experience to recognize as exaggerations, or even impossibilities, some of the reports they received.

Critical assessment of facts, particularly biological facts, is a comparatively recent habit; well into the seventeenth century AD, biologists still talked about the unicorn. They discussed the phoenix that lived for six centuries or so and was then regenerated by burning itself in a fire and rising, renewed, from the ashes. There was confusion, too, between animal and vegetable; there were supposed to be plants that produced animals – such as the tree that had barnacle geese as its fruit, or the one that gave birth to lambs at the tips of its branches. It has needed generations of careful observation and a wealth of detailed study into the behaviour of plants and animals to sort the true from the false, the fabulous from the unusual. Only in the light of centuries of experience, advanced recording techniques and the development of inter-disciplinary subjects like biochemistry and biophysics, has modern man been able to stride ahead of the biologists of the ancient world. Yet even now he is frequently astonished at what was done, what was known, and what knowledge has been lost.

Some of the earliest examples of careful biological observations are to be seen in the paleolithic cave paintings in Maux on the Ariège in the south of France. These show bison with arrows embedded in their hearts, and the motivation behind the art was most probably magical; paleolithic man lived by feeding on the flesh of the animals he could kill, and it was a common belief that by drawing what you wanted to happen you could actually influence the course of events. By painting bison with an arrow in the heart – which experience had taught them gave the easiest kill – they brought the bison under man's

Even in palaeolithic times, man was making careful biological observations, particularly of creatures he hunted. Cave paintings – such as this one of a bison at Altamira in Spain – reveal a highly developed understanding of the natural world. The artist has captured superbly the animal's shape and movement.

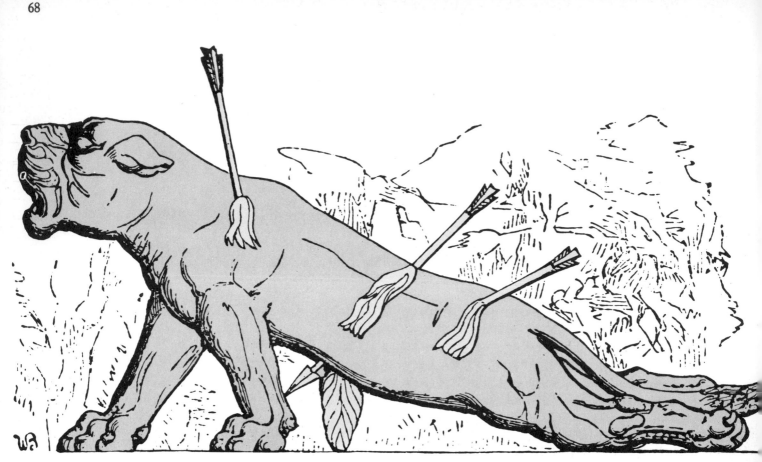

power through magic. But our interest today is both artistic and scientific, for the detail of observation shown is a remarkable tribute to the hunter's eye for shape and movement. And these cave paintings are typical of many of the period. They show, clearly enough, that even if motivated by magical considerations, the powers of observation of early man, and his ability to depict accurately what he had seen, was highly developed.

With this in mind we can turn to some remarkable and unusual Assyrian bas-reliefs from the ninth century BC. Cut in limestone and, in their day, painted black, white, blue, green and red, the ones we are concerned with show a date palm and, on each side of it, a divine being with the body of a man and the head of an eagle. These are symbolic pictures. The date palm is shown in a formalized way, with the trunk and the branches laid out in a rigid pattern; there is no beautiful, detailed, representational picture here. Yet we know that the artists of the time could depict scenes realistically, and there are plenty of other bas-reliefs that bear witness to their almost photographic accuracy. The famous carving of the hunted lioness is a typical case in point; here the animal is shown with three arrows sticking into it, and one of these has pierced right through the spine, causing paralysis of the hindquarters. Clearly the artist was drawing a scene he had observed, and with an accuracy of anatomical detail that leaves nothing to the imagination. So, to return to our symbolic date palm, it is evident that the symbolism was not due either to inadequate technique or to poor powers of observation. The carving was formalized because it was depicting a ritual act, and a closer examination shows what that was.

The palm stands between the two figures, each of which is holding out a hand as if offering a gift to the tree. And the gifts are nothing more nor less than the male flowers of the date palm. It is obvious then that what we have here is a fertilization symbol. From this it would seem that the Assyrians, and most likely their predecessors the Babylonians, knew of the sexuality of plants, although it is generally believed that this was not discovered until two and a half millennia

Assyrian bas-reliefs show keen observation
and a level of biological knowledge in
some areas not matched in the West for
thousands of years. **left** The famous picture of
the wounded lioness, for example, quite
clearly shows the paralysis resulting from
injury to the spinal cord. It has the immediacy
of a photograph.
bottom At first sight, this relief of
supernatural figures with a date palm
appears formalized and purely decorative.
But the figures are offering male flowers
to the tree. Clearly, the Assyrians
understood the sexuality of plants.

later. Is it possible? Perhaps one isolated case might leave doubts in our minds, but such Assyrian fertility symbols were not confined to date palms; some formalized bas-reliefs show fir trees, and mystical beings holding out fir cones. It would seem then that the Assyrians were in fact aware of plant sexuality; they did know that in some species there were separate male and female plants, and so fertilization from plant to plant was necessary. Yet this important fact was forgotten and had to be independently re-discovered.

Accurate observation was not the prerogative of the Babylonians; the Egyptians were equally adept, and the Greeks as well, although their artists were more concerned with depicting animals than plants. But Greek biology really centres round two figures, Aristotle and his younger contemporary Theophrastos. They make an interesting yet contrasting pair since Aristotle's preoccupation with living things centred around animals, not plants, while Theophrastos excelled in botany and took far less interest in the animal kingdom.

Theophrastos (not to be confused with Theophrastus von Hohenheim, the sixteenth-century AD physician who is known also as Paracelsus) was born at Eresos on the island of Lesbos, off the west coast of Turkey, sometime about 372 BC. He died in Athens in his eighties. Whatever the connotation Lesbos brings to mind today, it was primarily famous in Greek times for its early flowering of lyric poetry, and as the birthplace of many a famous philosopher, including Pittakos, one of Greece's 'Seven Wise Men'. Theophrastos was brought up there, but when a young man he moved to Athens to sit at the feet of Plato at the Academy. It was probably here that he first

met and became friendly with Aristotle. For a time Aristotle lived on Lesbos and perhaps, on vacations, the two men went out on nature study excursions together in the island. They probably walked around the shores as well, for Aristotle, certainly, had a predilection for marine life, and often mentioned the island lagoon at Pyrrha.

How friendly they became can be judged from the fact that when Aristotle was forced to leave Athens in 323 BC because of trumped-up charges of impiety, he gave the Lyceum that he had founded a dozen years before into Theophrastos's charge, and bequeathed him his extensive library and all the manuscripts of his own works. Here Theophrastos carried on his friend's traditions, and here he remained in charge for the next thirty-five years, enlarging and, to some extent, reorganizing it. And it seems that Theophrastos was not only a good administrator, but also a brilliant lecturer. He attracted a great number of pupils and – again like Aristotle – amassed an encyclopedic knowledge. At the end of his life he bemoaned the fact that he must die just when he was beginning to understand some of the mysteries of nature – the prerogative of every great man who is full of learning and retains his faculties to the very end.

Theophrastos is said to have written 227 treatises, ranging from medical matters and biological subjects to a book called *The Characters*, which became very popular. This was a series of some thirty sketches of typical weaknesses in human beings – arrogance, boorishness, buffoonery, superstition, and so on – subjects that others, including Plato and Aristotle, had written about before him. But Theophrastos was the first to collect a series together and, in doing so, started a literary style. However, we are concerned not so much with Theophrastos's wider interests as with his botany and, in particular, his astonishing system of plant classification. Through this he was able, single-handed, to lay the foundations of the whole of scientific botany. In two books, the *Enquiry into Plants* and the *Causes of Plants* he explained and showed how to distinguish the external organs, and invented the method of naming these in a regular manner, from the root upwards. He then extended this so that he effectively laid down the rudiments of all later botanical nomenclature; it is to Theophrastos, for example, that we owe the terms monocotyledon and dicotyledon for the two principal types of plant seeds. He also appreciated that there was a relationship between the structure and function of a plant, and was the first to discuss the question of geographical distribution. And having this as a basis, he was able to make a classification of all plants, a classification so good

Further evidence of the Assyrians' superb delineation of botanical detail : Part of a relief from the royal park at Nineveh showing vines and flowers.

that it stood the test of time and was not materially improved until Carl Linné (Linnaeus) founded our modern classification system in the eighteenth century AD, more than two millennia later.

This long delay, this virtual loss of Theophrastos's work, was due not to the destruction of his manuscript, but to the fact that scientific botany made no progress after him – a result of the Latin world's emphasis on the medicinal use of plants, rather than on their nature *per se*. The West took no interest in what Theophrastos had found until the thirteenth century, when a compilation of Aristotle's biology and Theophrastos's botany became available; yet even then the compilation was fathered solely on Aristotle, and no one realized what Theophrastos had contributed. Only late in the fifteenth century did the situation change, after the first translation into Latin of the *Enquiry into Plants* had appeared, correctly attributed. Yet if Theophrastos's work had been recognized, a whole side of botany would have been known, a side that was totally and completely neglected so that any scientific approach to plant life and plant structure was stifled before it could begin.

Theophrastos, like the Babylonians, also knew of the sexual nature of some varieties. In the *Causes of Plants* he describes the fertilization of the date palm, saying quite specifically that 'it is helpful to bring the male to the female', and going on to describe the action of each part. Yet his knowledge was of no more use to later ages than that of the Babylonians had been; scientific botany had to wait for this vital fact to be rediscovered.

Theophrastos has been called the 'Father of Botany'; if this is a valid tag – and there is every reason to think it is – then it is equally appropriate to say that Aristotle was the 'Father of Biology'. He stood alone in the ancient world, and remained head and shoulders above any other biologist for the next two thousand years. Yet it has been one of fate's many ironies that it is not for his biology that Aristotle has been mainly remembered, but for his writings on ethics, on politics, on rhetoric and philosophy, on physics and astronomy, and even on metaphysics; in fact for everything but his biology.

Aristotle's biological writings were not lost, and his powers of detailed observation were sometimes remarked on, but much of his natural history seems to have been looked upon more as an indulgence that had given him pleasure than as a serious scientific study. It was readily admitted that his work on bees and their diseases was useful, especially since honey was the primary sweetening agent until comparatively recently, but there were always those who were ready to point out his mistakes. And none were more willing to do this than the Renaissance biologists. Only a fresh spate of biological observing and theorizing in the nineteenth century AD brought to light the fact that many of the errors attributed to him were not errors at all, and led to a realization that Aristotle the biologist had been grossly underestimated compared with Aristotle the philosopher.

Theophrastos has a good claim to the title 'Father of Botany'. **below** This apocryphal portrait of him is from the *Nuremberg Chronicle* of 1493, while the two plants shown here are from one of the earliest printed herbals, published by Joanne Ruellio in 1478. This was a copy of the herbal of Dioskorides compiled in the first century BC. **left** The lady's slipper orchid, *Cyprepedium calceolus*, belongs to the group of plants classified by the structure of their seeds as monocotyledons. **right** The celery-leaved crowsfoot, *Ranunculus scleratus*, however, is a dicotyledon. Theophrastos was the first to distinguish between these two types. He drew up a comprehensive classification of plants, and made many other botanical discoveries.

As an observer of the animal world Aristotle's forte lay in ichthyology, the study of fishes, and yet it was on some of his descriptions of marine life that he was later to be criticized. For instance, in talking about squids, octopods and similar creatures – the cephalopods – he described the nature of one of the tentacles of the male and how it was used to fertilize the female. Yet later generations scorned the idea because Aristotle's description was of a specific octopod – the paper nautilus, or argonaut – and only female argonauts were known. Only after fresh and more detailed observations were made between 1827 and 1842 was it appreciated that Aristotle had been right after all.

Again, in his studies of catfish, Aristotle claimed that during breeding it was the male, not the female, that watched over the young for a period of fifty days, and also that in the Achelous river that runs into the Gulf of Corinth, these fish made a kind of squeaking noise by rubbing their gills together. But he was disbelieved, not least because western European catfish do not behave like this during breeding, and it was not until 1856 that the naturalist Louis Agassiz at Harvard found that North American catfish do just as Aristotle described. He then had some catfish from the Achelous sent over to him, and was able to verify Aristotle's account, leaving no room for doubt – although the confirmation he had made did not become common knowledge in the scientific world until fifty years later. As far as squeaking catfish, and some other fish, are concerned, once again Aristotle was right; although to be precise it is the gill cases rather than the gills themselves that rub together.

Aristotle's powers of observation were tremendous, but to get the results he did, he must have spent an immense amount of time, peering from a boat into the clear sunlit waters in the Pyrrha lagoon, or on the Achelous. And although this doubtless brought him intense delight, he never neglected to make notes and prepare detailed descriptions of what he saw. Of course, there is little point in going on to describe all the many observations he made of marine life, let alone of land animals as well, but there was one more discovery, later to be ridiculed and ignored, that must be mentioned. This concerns the dogfish that he called 'galeos', which 'lays eggs within itself'. Aristotle knew that, as a general rule, fishes bring forth their young as eggs, not as small, independent, moving creatures as mammals do. He was, however, well aware that a few marine animals resemble the mammals in bearing live young; these he classified as 'selache'. (The modern term selachian refers to all sharks, rays and dogfish.) Of the selache, one species was very special, he said; this was the galeos, a placental dogfish.

Aristotle knew that when a mammal bears young, the embryo is attached to the mother's womb by a cord, and that this point of attachment is for ever to be seen in the fully grown animal as the navel. At the other end of the cord there is a large, flat, fleshy organ that is ejected by the mother after the offspring has been born – the

above Aristotle's catfish, *Parasilurus aristotelis*: Aristotle observed that it was the male of the species, not the female, that guarded the eggs until the young hatched out. He was disbelieved until, in the middle of the last century, Louis Agassiz discovered that the North American catfish, *Ameiurus*, behaved in exactly the same way.

right Aristotle's 'galeos', the placental dogfish, reproduces rather like a mammal. It bears live young, and the embryo (as shown here) is attached to the inside of the mother's body by means of a placenta. This Aristotle discovered and described, but his description was dismissed as fantasy. It was only in 1852 that Johannes Müller proved the truth of what Aristotle claimed.

below This beautiful mosaic of marine life is in the Ancient Roman town of Uttica, in present-day Tunisia. It has the clear immediacy of direct observation, and a number of Mediterranean species can be recognized clearly.

Aristotle's organization of the natural world into a scale of complexity was based on a concept we no longer accept : the relative presence of three types of 'soul'. Nor did Aristotle imply any development in the evolutionary sense. But that does not detract from the importance of this classification, the first to attempt to embrace everything from inanimate matter, through plants and animals, to man.

so-called 'afterbirth', or placenta. Now, what Aristotle said about the placental dogfish (which he described, naturally enough, as 'a shark-like fish', for that is what it looks like) was that the eggs were deposited in the womb of the mother, and that the young develop with a navel-string attached to the wall of the womb. He described how this attachment was by way of a sucker (the placenta), and how each embryo had separate embryonic membranes just as mammals do. The whole detailed description was astonishing to anyone at all familiar with marine life, and knowledgeable about selachians, and it is perhaps not surprising that, although it was read, it was passed over as too fantastic to be true. Not until 1852 did the naturalist Johannes Müller show that Aristotle had been quite right, and that the galeos is one of a very select few fishes that resemble mammals in the way they bear their young.

Aristotle did not restrict himself to describing animals, marine or otherwise; he was a philosopher and could not resist thinking and writing about the wider implications of what he had observed. Unfortunately the full import of what he had to say was, like the rest of his biological work, largely ignored and forgotten, and the use it could have been in stimulating biology was not realized until the sixteenth century AD. Yet it is to Aristotle that we owe the first great classification of animals. A total of 540 different species were described, and it is worth noting that this classification was no mere series of lists with brief comments; he stressed the importance of learning the different parts by dissection, and he himself dissected anything between one and ten different animals of each type he classified. It was this work that led him, for example, to describe in detail the four chambers in the compound stomach of ruminants, such as cows – a fact that no one else mentioned until Volcher Coiter did so sixteen hundred years later.

The classification Aristotle devised would not do today, for it was based on the concept that there were three different types of 'soul': the nutritive or vegetative soul that gave capacity for nourishment and growth, the sensitive or animal soul which provided sensibility to external conditions, and the intellectual or rational soul that was to be found only in man. But the fact that we now adopt different, and more physical, criteria for classification in no way lessens the significance of what Aristotle did. His scheme embraced all nature, organic and inorganic. Inorganic or non-living material was that which possessed no soul at all; the vegetable kingdom had a vegetative soul but no other; while animals contained both vegetative and animal souls. Like his astronomy and physics, it was a synthesis of the natural order. But it had wider implications, and led Aristotle on to his scale of nature.

This was a brilliant application of his classification by behaviour – which is what his classification by soul really was – and is essentially a scale from the most simple to the most complex. As Aristotle himself put it: 'Nature proceeds little by little from things that are lifeless

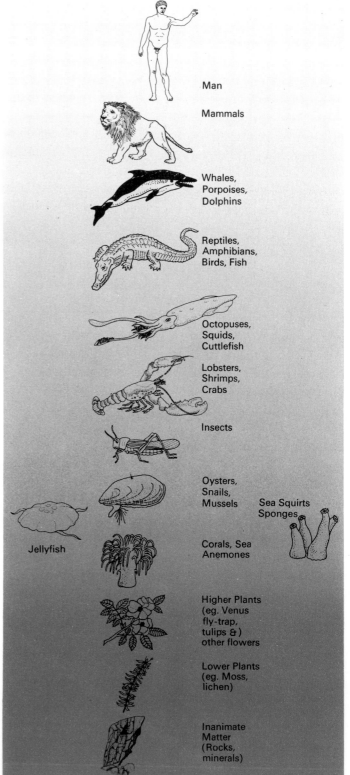

to animal life in such a way that it is impossible to determine the exact line of demarcation . . . Thus, next after lifeless things in the upward scale comes the plant, and of plants one will differ from another as to its amount of apparent vitality.' And so he drew up his scale, beginning at the bottom with inorganic inanimate matter, then moving upwards by way of lower plants to those with greater 'apparent vitality' and on to animals, starting with those that seem simple and almost plant-like, such as sponges, and proceeding by stages of increasing complexity to reach mammals and then man.

Did Aristotle consider this in an evolutionary sense? If we are thinking of fully-fledged Darwinian evolution by natural selection, then the answer must be no; but he did consider that in each animal the various parts were organized and united for its greatest good. The step he did not take was to consider this organization between different classes in his scale, except that clearly he recognised a development in complexity of behavioural characteristics. Could one arise from the other? Aristotle does not seem to have asked this specific question, perhaps because he was concerned so much with the different forms and their perpetuation from parent to offspring; but his scale left it open. Someone could have asked it – and possibly have brought about a dynamic evolutionary view some two millennia earlier – yet no one did. And this is in spite of several earlier ideas that might have contributed to such a development.

Anaximander, in keeping with his belief that water was the fundamental natural element, suggested in the sixth century BC that man and animals had developed from fish. In the opinion of Empedokles a hundred years later, the limbs of animals had been assembled into various creatures, of which only those best fitted for life had survived. Indeed, the Greek view that fossils were the remains of animals that no longer existed could also have acted as a stimulant. But, unfortunately, the Muslim world and the medieval West ossified his views and a new spirit of independent enquiry had to be reborn before the next step could be taken. Aristotle's scale of nature may not be a lost discovery of evolution that had to be found again by Charles Darwin, but it was a conceptual framework that, coupled with Aristotle's detailed work on living organisms, could have provided the springboard for such an idea. That it did not is the more surprising in view of what Lucretius was later to say in *De Rerum Natura*.

Lucretius's epic poem is concerned with the entire natural world, organic as well as inanimate, and in the section in which he wrote about the coming of animals we find him a step beyond Aristotle, suggesting not a specific developmental evolution, but a series of attempts by nature to produce all kinds of creatures: 'In those days the Earth attempted also to produce a host of monsters, grotesque in build and aspect . . .' But none of these flourished, they were 'all in vain', and Lucretius gives the reason: 'In those days, again, many species must have died out altogether and failed to reproduce their

kind. Every species that you now see drawing the breath of life has been protected and preserved from the beginning of the world either by cunning or by prowess or by speed.' A succinct and clear description of what, after Herbert Spencer had coined the phrase, became known as the survival of the fittest.

Here, then, was a view that, coupled with Aristotle's scale of nature and a knowledge of fossils, might well have led to the idea of a truly dynamic evolution. But it did not. The embryonic concept was lost in the static belief of a fixity of species and the mistaken notion that they were all created at once. This approach fitted in well with the biblical outlook and the general medieval synthesis of man, creation and the deity, but it stifled any other view. How entrenched it remained was evident when Darwin, Huxley and Spencer revived these views in the 1860s as part of the full theory of evolution.

CHAPTER 6

Medicine, Drugs and Psychology

MAN has always collected and collated remedies for sickness. Gradually, in the fullness of time, he learnt which ingredients could help him and which could not, which were effective specifics for different ills and which were dangerous or deadly. Side by side with this simple medicine, he acquired a growing body of knowledge about the human system, gleaned partly from parallels with animals butchered for food, and partly from tending the wounds of hunters or soldiers; indeed, in late Roman times, in the second century AD, the surgeon Galen was still learning much about the human body by attending gladiators injured in combat. Early medicine and surgery was primarily a matter of trial and error. It was bound to be; successes would be remembered and imitated, mistakes would be forgotten and, gradually, over very long periods of time, a corpus of knowledge and a code of practice would be built up. And although this sounds haphazard enough – as it doubtless was – it is surprising how many pre-echoes of modern treatment we find, and how much simple surgery was undertaken using herbal anaesthesia.

There were echoes, too, of modern medical ethics. Sometime in the late nineteenth or early eighteenth centuries BC, King Hammurabi of Babylonia enacted his code of laws. This contained very stringent rules for surgeons, including how much they should be paid, the fees varying according to the social standing of the patient, and what penalties should be extracted for failure. These were the more severe the nobler the patient; an eye for an eye and a tooth for a tooth, metaphorically if not literally. Physicians received no mention; perhaps they formed a separate semi-sacred élite, not amenable to ordinary laws.

The most famous code of conduct, though, and one that in its various forms has influenced medical practice up until the present day, is that based on a short text by the Greek medical genius, Hippocrates of Cos. It has become known as the Hippocratic Oath. Although essentially a set of guild rules binding the members of the profession together, it is now often confused with another work, the Greek medical *Law*. This was probably of later origin, but nevertheless enshrined a code that may well have been earlier practice. Between them, the *Oath* and the *Law* set out a standard of medical behaviour that has been followed in every Western country, and has commended itself virtually everywhere else. This is primarily because the art of medicine has always attracted a certain kind of individual, whose aims have been the same – to keep the functions of the body

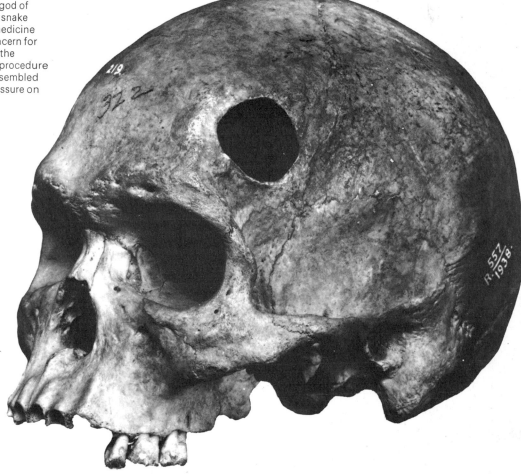

left A statue of the Graeco-Roman god of healing, Asklepios, with the sacred snake twining up his staff—a symbol of medicine that emphasizes man's age-old concern for health. **right** A trephined skull from the island of New Britain. This ancient procedure was largely religious, but closely resembled the modern method of relieving pressure on the brain.

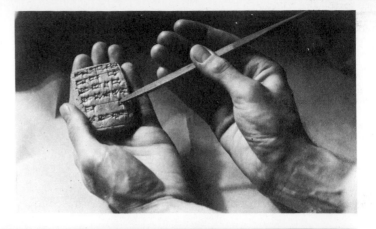

below This large cuneiform tablet, found at Nippur, has been called a Mesopotamian 'dictionary of medicine'. It dates from about 1300 BC, and describes treatments for various ailments.
right Writing on a clay tablet, using a wedge-shaped stylus: The shape of the writing implement made it natural to form cuneiform rather than curved inscriptions.

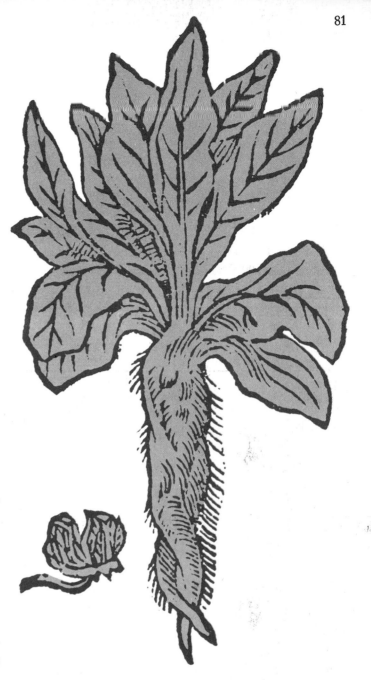

The mandrake plant was used extensively in the eastern Mediterranean region in ancient times as a medicinal herb. In large doses it acted as an anaesthetic and in weaker ones as an aphrodisiac. The illustration here is from an edition of the herbal of Dioscorides (originally written in the first century AD) printed in Lyons in 1478.

in balance, to see that everything is working well so that the whole human organism is in harmony; to expel, to exorcise every ill; in short to promote good health. But if the goal has been a common one, the methods have varied widely and, unfortunately, much that was efficacious has been forgotten and required rediscovery, sometimes in another form.

Anaesthesia is one of those discoveries that were, no doubt, an extension of experience. In the ancient world this usually took the form of a soporific drink. The Egyptians used nepenthes, a plant extract of some kind which acted as a strong sedative. (The name is not Egyptian but a later Greek word describing its action and meaning, literally, 'no grief'.) There is evidence that they also knew of the mandrake. This plant, which could be used as an anaesthetic or, in weaker doses, as an aphrodisiac, was familiar to the Babylonians, the Hebrews and the Greeks, as well as the Egyptians, but was difficult to handle because of its side effects. In medieval times a mixture of various soporifics including opium was used, but one drug of this kind that is only now being replaced, yet whose discovery was lost for nearly three thousand years, was cocaine. Used in the first millennium BC by the Incas of Peru, partly as an addictive drug, its properties were also known and applied for medical reasons, especially when amputations were being undertaken. The Incas did not perform any chemical process to extract cocaine from the coca plant; they used the leaves direct. But their practice was only learned of by Europeans in the sixteenth century AD, and even then it was not applied in Western medicine until the 1880s.

But if the use of anaesthesia was widespread and continuous in most early civilizations, it was forgotten later in the Western world. As a result, sterner methods were sometimes used, at least by the military, although the general consequence was that surgeons were limited to operations that could be performed in just a few moments. This often resulted in the most astonishing speeds; the famous eighteenth-century English surgeon William Cheselden could do a lithotomy (removal of bladder stones) in fifty-two seconds, and even as late as 1821 Charles Bell, a well-known Scottish surgeon, could still write that an amputation of an arm at the shoulder 'requires decision and rapidity; and the knife is to be handled more like a sabre than a surgeon's scalpel'.

This retrogression due to a loss of earlier medical knowledge was no isolated case; there were other ideas that should have been remembered but were not. The Egyptians, ever practical, encouraged doctors to specialize, and thus developed quite sophisticated techniques, particularly in dentistry. The well-to-do of Ancient Egypt might have their teeth stopped with gold, and both rich and poor could get their dental abscesses drained. (The high incidence of arteriosclerosis – hardening of the arteries – may have caused this to become a comparatively frequent operation.) Mouldy bread was applied to wounds – an early example of the use of antibiotics.

Animal excrement and mud were also concocted with other ingredients into medicines, thus making available, we now recognize, bactericides and substances like the antibiotic aureomycin, which is a natural constituent of Nile mud.

One of the most surprising things, when one comes to think about it, is the fact that as early as the middle of the third millennium BC, when the great pyramid of Gizeh was being built and there were labour camps crowded with a multitude of workmen – Herodotos, writing a millennium later, claimed that a grand total of one hundred thousand were employed – there were no disastrous epidemics. Considering the crowded and primitive conditions in the camps, the lack of modern methods of hygiene and the high everyday temperatures in the area, one would expect illness to spread like wildfire. Yet it did not. The reason for this astonishing bill of good health only became apparent in 1948. The men's diet contained large quantities of radishes, garlic and onions, and it was realized that here was a series of natural inhibitors of bacteria, effective against typhoid, cholera and dysentery. This does not mean these diseases were totally absent or that there was no sickness at all, but that the diet was one that materially reduced the likelihood of large-scale infection. What did the Egyptians know of this? Clearly not that they had antibiotics or bactericides; but they grasped the purely practical fact that

Much ancient medical knowledge was lost in medieval times. **below** This hideous example of the primitive methods used dates from 1593. A man's leg is amputated without any anaesthetic except the power of the fist of the surgeon's assistant (seen on the right). **opposite** A tomb painting at Beni Hassan in Egypt dating from about 2000 BC shows an acacia tree. The Egyptians made a chemical contraceptive by grinding together acacia spikes, honey and dates. We now know that acacia spikes contain lactic acid, a chemical that kills sperm.

The mud of the Nile, dried hard by the Sun after the annual flooding has subsided. The Ancient Egyptians used this mud as an ingredient in a number of medical preparations, and only recently it has been found to contain natural antibiotics.

some foods were necessary in quantity if the health of a crowded working population was to be maintained – a fact culled, no doubt, from the experience of countless numbers of previous generations.

On matters of sex and child bearing, the Egyptians had a number of advanced ideas that never reached the Western world. For instance, they used the equivalent of a contraceptive jelly which was applied on a wad of fibres inserted deep into the vagina. The mixture was composed of honey, dates and acacia spikes ground to a very fine consistency. Not until many millennia later was the West to learn of this, and only in recent years has it been found that the method was effective because acacia spikes contain a gum that proves deadly to sperm. The gum's active constituent, which is released when it is dissolved in a fluid, is lactic acid, a component of many contraceptive jellies today.

Again, in Ancient Egypt, it was possible to have a pregnancy test at the earliest stages and, at the same time, to determine the sex of the unborn child – or so it was claimed. The method they used was to take a woman's urine and soak bags containing wheat and barley with it. They found that if the subject was pregnant, the urine would accelerate the growth of the wheat if the child was to be a boy, or the barley if it was to be a girl. Yet such a test is only a comparatively recent innovation with us; not until 1926 was a urine pregnancy test discovered, and it was another seven years before the acceleration of wheat and barley was confirmed by laboratory tests.

A certain amount of surgery was practised in Ancient Egypt, but there were no operations whose techniques were lost and had to be rediscovered in later ages. Presumably the Egyptian surgeons gained their knowledge as every other early surgeon did, by trial and error, although it is conceivable that the embalmers passed on something. Mummifying animals and humans seems to have been carried out from the earliest days of the Egyptian state, sometime in the third millennium BC, or perhaps even earlier. In practising their art, embalmers became familiar with the organs, the cavities and other aspects of the internal anatomy of the human body. Since embalming was a prelude to funerary rites (or possibly part of them), and the priesthood were intimately concerned with healing, a link between surgeons and the embalmers is not unlikely.

Another amazing fact is that the Egyptians who went on the desert caravan routes across the Sahara westwards to the Kharga Oasis, and south-west to Lake Chad, chewed a root that they called ami-majos. They found that this gave them extra protection from the Sun by reinforcing their skin pigmentation, and modern research shows that the root contains the active organic chemical compound called 8-methoxypsorate.

The Egyptians also practised psychological as well as physical medicine, at least in the Late Period of indigenous Egyptian rule, which means sometime later than 1100 BC. The treatment was linked with the priesthood – who have, in all ages and in all religions, offered counsel and advice, acted as mediators of divine comfort and consolation, soothed as well as exhorted the faithful. In this sense what the Egyptians did was no different from what seems always to have been done as part of the pastoral duty of the priest, but in another sense what occurred in Egypt was unusual. There the priests used drugs as well as advice, potions as well as prayers; they induced sleep and interpreted dreams. This treatment was linked with worship of Imhotep, the god of healing, and the patients attended specially built healing temples, particularly those at Memphis and Saqqara. The drugs used were mainly henbane and opium, although some healing brews were administered too, and it was suggested to the patients that in sleep they might meet the god in person and be cured. Some doubtless thought they had, and believed themselves cured – and perhaps they were.

The Egyptians amassed a host of medical facts but, as far as we know, never drew up an organized corpus of medical knowledge. (The medical papyri such as the Ebers Papyrus and Smith Papyrus were merely collections of remedies and practical instructions.) In this there was a parallel between them and the Babylonians, for their medical treatments were as practical as those of the Egyptians

but just as lacking in formalization. The first real systemization of medical knowledge appears to have come with the Greeks. How much of their earliest practice they may have derived from the Egyptians and Babylonians is uncertain; however we do know that they too practised psychological healing, only their healing temples were dedicated to their own medical god Asklepios. Indeed, with the Greeks, the whole practice was widespread; there were some 320 centres for treatment, which included purifying baths and interpretation of the dreams experienced during the period of induced sleep.

The Greek Hippocratic tradition grew up in the fourth century BC. Hippocrates himself was born on the island of Cos sometime about 460 BC, and lived until he was eighty-five. But, like Pythagoras, he had a great following, and it is now impossible to sort out what is his original work and what is due to members of his 'school'. Today all their writings, the whole output of Hippocrates and his followers, are classed together as the Hippocratic Corpus. One thing about which there is no doubt, though, is that Hippocrates was outstanding as a physician and deserves his legendary fame. While he was perhaps the greatest, and is certainly the best remembered, of the early Greek

physicians, and while his reputation has swallowed the individuality of members of his group, there were other important medical practitioners whose names have come down to us. Among them is the philosopher, physician and scientist Alkmaion, who was a disciple of Pythagoras, and was born around 500 BC at Croton.

Alkmaion's importance lies in two discoveries he made. The first was his realization that the brain is the centre of all sensations. (Generally in the ancient world the heart was given this place of preeminence.) We can only conjecture how he came to this conclusion, but it could well have been from direct evidence. Any practising physician or surgeon would see head injuries and perhaps the resulting paralysis, and during trephination, which was not an uncommon operation, mistakes could be made and the brain damaged. So far as evidence is concerned, then, he had no more material to work on than his contemporaries, but he was a meticulous observer. It may well have been a combination of observation and inspiration that led him to his important conclusion. Alkmaion's faculty for observation must also have helped him in his dissections, and we can make an informed guess that this was how he came to be the first to describe the channels that run from the throat to the middle ear. These channels we call the Eustachian tubes in honour of the sixteenth-century AD anatomist Bartolomeo Eustachio who was later hailed as their discoverer.

Yet another Greek philosopher whose pioneering medical work was to be confined to oblivion for generations was Empedokles. Born in Agrigentum (modern Agrigento) in Sicily in the fifth century BC, he was that Hellenistic rarity, a practical man as well as a philosopher. He had an immense reputation in his own day as a philosopher, orator, statesman, social reformer, doctor, and miracle-worker. Although it is difficult for us to sort fact from fable, one thing is clear: He was convinced that the constant epidemics of malaria in Agrigentum were caused by an infection from the near-by marshes, and he had them drained at his own expense. The result was, as we should expect, a marked drop in the incidence of the disease, but his lead was not followed elsewhere, perhaps because he could give no exact reason for his success. Did Empedokles recognize that the mosquito was the carrier of the disease? Or did he think it due to some kind of emanation from the marshes? One's immediate reaction is to favour the latter, until one realizes that the scholar and librarian Marcus Varro, in the first century BC, claimed that malaria was carried not by vapours but by 'small animals' (bestiolae) from swamps. Clearly he was echoing some earlier view, since he himself carried out no independent research, medical or otherwise. If Empedokles did realize that the mosquito was the carrier, and if his efforts had been imitated, how much sooner would this debilitating scourge have come under control.

Our main source of late Greek medical knowledge is Alexandria. It blossomed into prominence as Athens declined early in the third

86

bottom Embalming the dead in Ancient Egypt must have led to many anatomical discoveries, and this information may well have reached the healing priests, who probably worked closely with the embalmers.

century BC, and a considerable amount of medical research was carried out there, research which presaged all kinds of work not to be repeated until well after the Renaissance. In the years following its foundation, Alexandria could boast of two outstanding medical men, Erasistratos and Herophilos. True, they have been accused of using criminals for human vivisection but the evidence against them is meagre; what we do know for certain, though, is that the dissection of dead bodies was common practice at the time.

Herophilos, born at Chalcedon at the close of the fourth century BC, moved to Alexandria early in the following century, and has been called the Father of Anatomy. He was certainly the founder of systematic anatomy, and if anyone deserves the title, it is him. He took a particular interest in the reproductive organs and the brain. As far as the brain is concerned, he described various parts, including the place where the veins meet at the back of the skull, still known as the *torcular Herophili*. He also described the *rete mirabile*, the web of blood vessels found at the base of the brain in sheep but not in man,

although Herophilos did not realize this was so. Obviously, then, animals were being used for dissection, and we have no clear indication of what anatomical information he obtained direct from the human body and what he gathered from animals and assumed to be the same in man.

From his work on the brain, Herophilos was led to examine and consider the nerves, and he was the first to distinguish between the motor nerves, which excite muscular activity, and the sensory nerves, which carry messages from sense organs to the brain. This distinction was not re-introduced until the eighteenth century AD. Again, on the reproductive organs he did work that lay dormant for almost two thousand years. He discovered and described the ovaries and, above all, he found the thin tubes that run from them to the womb. These were the fallopian tubes, whose name gives credit to Gabrielle Fallopio who found them again in 1561.

Our other great Alexandrian doctor, Erasistratos, was born in Iulius, the chief town of the island of Keos (modern Zea) and for a

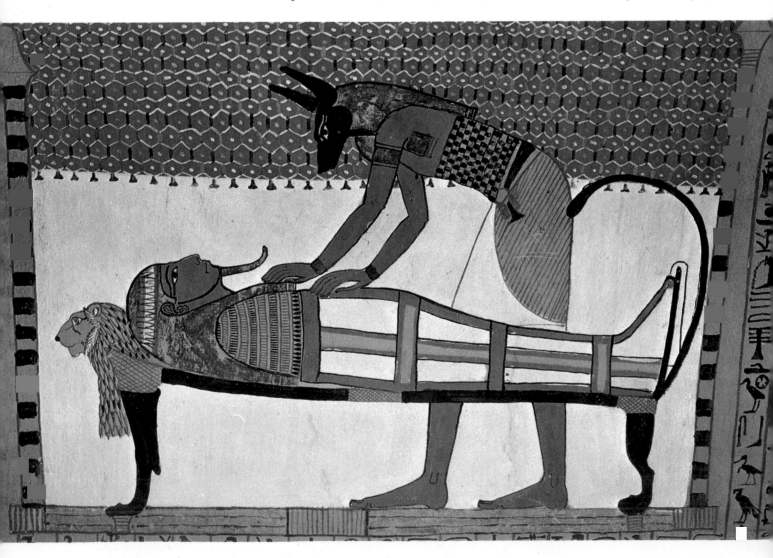

bottom A fragment of an Ancient Egyptian medical text, thc London Medical Papyrus. This part deals with women's ailments.

above A vase painting shows a Greek healing temple in the time of Hippocrates. The physician in the centre is bleeding a patient. Some 320 such medical centres were established in Greece; they offered both psychological and medical treatment.

on the cardiovascular system – the heart, the veins and the arteries.

Erasistratos weighed birds and animals before and after eating, and also discussed the changes that go on in the body during the breaking down and building up different substances – the body's metabolism. What he called the substances does not matter; what is significant is that he appreciated the essentials of metabolism, a subject that was afterwards neglected until the Italian Santorio Santorio revived it in the early seventeenth century AD. Erasistratos also explored the cardio-vascular system, and has been credited with discovering the circula-tion of the blood, pre-dating William Harvey's work that virtually coincided with Santorio's. This seems to be an exaggeration of what Erasistratos knew, for although he talked of the dilation of the arteries, he did not refer to any continuous flow around the body. He certainly recognized the heart as a pump – 'like a blacksmith's bel-lows' was his description – but he did not quite manage to find the complete answer. Had Erasistratos's immediate successors accepted his view, it could well have led them to recognize the circulation. As

time was Herophilos's pupil, although he only moved to Alexandria to do research there late in life. Just as Herophilos has been named the Father of Anatomy, so Erasistratos has been called the Father of Physiology, and with equal justification. Perhaps due to Herophilos's teaching, he too took a great interest in the nervous system, and traced the paths of the motor and sensory nerves in considerable detail. He also studied the digestive system and described the action of the gastric muscles. But the points of particular interest in his work that were not appreciated by those, like Galen, who read his books, were his basic ideas of what we now call metabolism, and his work

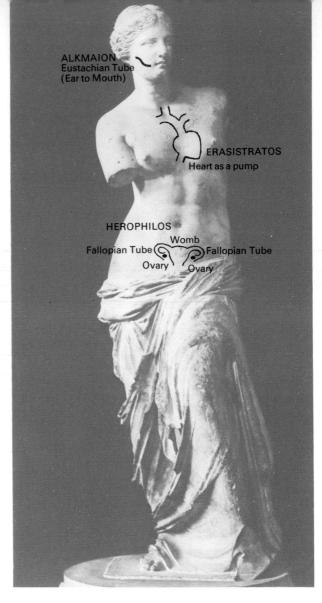

ALKMAION
Eustachian Tube
(Ear to Mouth)

ERASISTRATOS
Heart as a pump

HEROPHILOS
Womb
Fallopian Tube   Fallopian Tube
Ovary   Ovary

bottom The *Anopheles* mosquito is the carrier of malaria: was this the 'small animal' from the swamps that Marcus Varro blamed 2,000 years ago for spreading the disease?
right A number of remarkable discoveries in internal anatomy were made by Greek medical men in pre-Christian times, only to be lost for many centuries. Several of them are illustrated here.

it was, the old idea of the ebb and flow of blood within the body, a kind of tidal theory, continued to be accepted.

If only part of Greek medicine and medical knowledge from Egypt and Babylonia filtered through to the West, virtually nothing came from those civilizations that lay beyond the immediate boundaries of the Mediterranean area, except perhaps for the odd rumour of some miraculous cure. But now, with archaeological and historical research behind us, we are able to make some assessment of what was achieved, and what we find is often surprising. In India, for example, there were a whole host of herbal medicines. In the famous *Charaka Samhita*, a compilation of medical knowledge prepared in the first century AD by the court physician Charaka, there is a list of more than 500 herbal drugs, a pharmacopoeia more extensive even than any Mediterranean one. And while the majority are effective but unremarkable, the list does mention the Indian plant *Rauwolfia serpentina*. This is a form of snake-root, and its Latin name is derived from that of the sixteenth century German physician and botanist Leonhard Rauwolff, who drew attention to its properties as a sedative. Yet it had been in use at least fifteen hundred years earlier in India, where it was prescribed for colic, headaches and, above all, as an anti-depressant – the 'medicine of sad men' it was called. Modern chemical analysis has shown how right the Hindu physicians were, for the plant contains a number of powerful alkaloids including reserpine, a tranquillizing drug introduced into Western psychiatric medicine only in the 1950s.

Two interesting and unusual ideas also permeated early Indian medicine. In the first place they carried out plastic surgery. In the

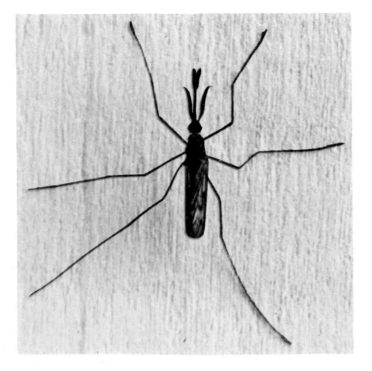

Susrutu Samhita, a collection compiled in the fifth century AD or possibly even earlier, and concerned mostly with surgery, the surgeon is instructed how to use skin from the cheek or from the forehead to replace a nose lost through disease, perhaps, or cut off by accident or as a punishment. The first recorded skin grafts in the Western world were carried out in the fifteenth century AD, more than a millennium after the practice was well established in India.

The second interesting point about Indian surgery was the use of sutures ('stitches') for binding the edges of surgical wounds. Again in the *Susrutu Samhita*, there are descriptions of the kind of curved needles, made of bone or bronze, which were not adopted in the West until the nineteenth century, and one ingenious method of suturing that never has been followed up. This was the use of the large black Bengali ants for joining intestinal wounds. The ants were placed side by side and would clamp the edges of a wound in their jaws; they were then decapitated and their bodies removed, the heads remaining behind to dissolve away by the time the wound had healed. The intestines, complete with their macabre suture, were then replaced and the abdomen sewn up.

If Indian medicine took a long time to reach Western Christendom, so too did knowledge about the medical practice of the South American civilizations. Yet here, too, there were novelties and unexpected advances. In Mexico, for example, ceramic figurines made most probably during the first millennium BC show evidence for caesarean births. They seem to point to the high incidence of the scourage of syphilis, a disease which Columbus's crew and Cortez's army brought back in a very virulent form to the West. It is to the

below A photograph of part of Trajan's
column shows wounded Roman legionaries
being treated on the battlefield. It was through
treating wounds – either military or gladiatorial
– that Galen and other Roman physicians
gained their basic knowledge of human
anatomy and physiology, for dissection of the
dead was banned throughout the Roman
empire.

below The title-page of the *Anathomia* of Mundinus, published in 1493, shows a demonstration of those parts of the body described by Galen. (The reader is referring to one of Galen's books.) The object was not to make new discoveries but to prove the truth of Galen's writings, and this epitomizes the attitudes of men in the Middle Ages and later.

They referred to the ancient authorities in matters of science just as they did to the Church elders and the Bible in matters of faith. There was no widespread spirit of independent inquiry, and knowledge of ancient science was in any case far from complete.

right The snakeroot, *Rauwolfia serpentina*, used for centuries in Indian medicine, is known to contain a potent tranquillizer.

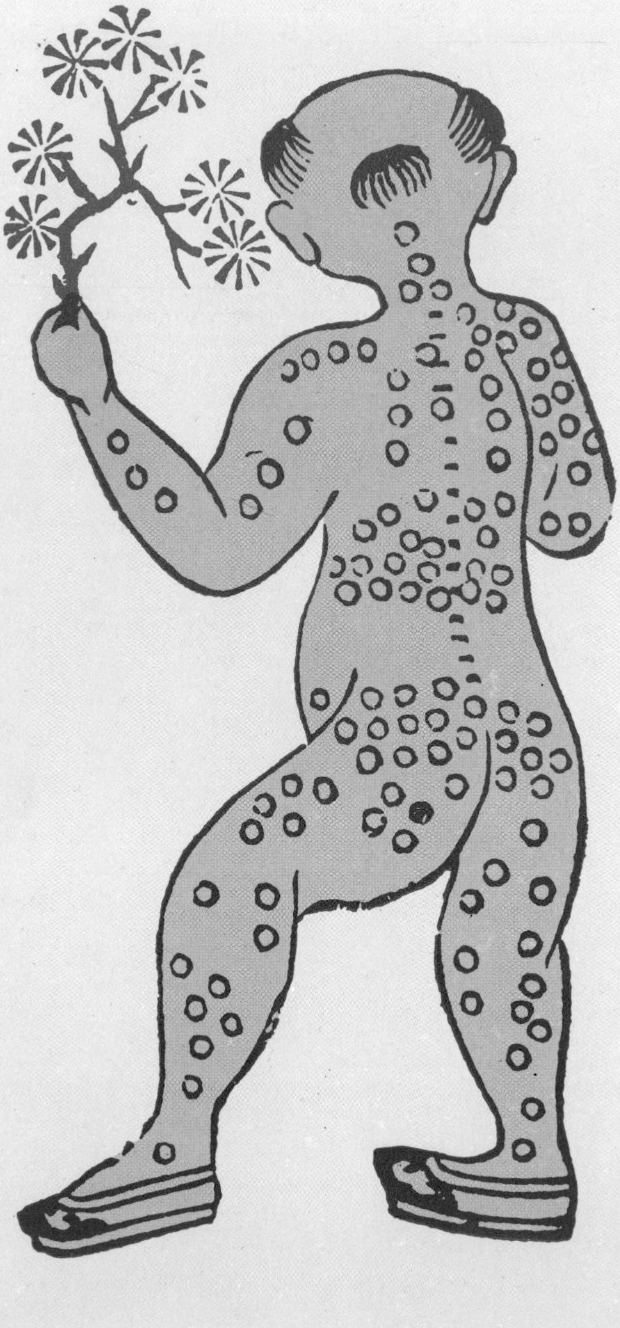

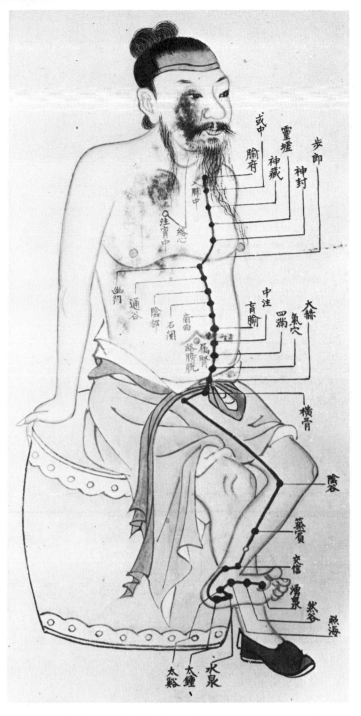

sought a balance between the two disparate qualities, the Yin and the Yang – the one cold, dark, female; the other warm, sunny and male – and all treatment was aimed at achieving this. But, just as in other civilizations, so in China, by trial and error, by practice and experience built up over long periods of time, the Chinese hit upon certain powerful drugs and treatments whose efficacy modern research has confirmed, and which would have been helpful to the Western physician had he only known about them.

The Chinese, like all other early civilizations, used herbal remedies, but they were not confined to them. They used minerals when they thought these would do good, and for syphilis they prescribed mercury, a treatment that was not used in the West until the sixteenth century AD, two thousand years later. For eye infections copper sulphate was used, as it was in Egypt, but another of their own innovations was sulphur for purifying the blood. Among their herbal remedies three were notable: *Ephedra sinica, Dichroa fibrifuga*, and chaulmoogra oil. The first is the Chinese joint-fir which they used in the treatment of coughs and lung ailments; in 1887 this was found to contain ephedrine, a powerful drug used now in cases of asthma and bronchitis, but only adopted in Europe in 1925. The second is a plant root adopted as a specific against malaria, yet only in 1948 was it found that the root does in fact contain alkaloids that are an effective treatment against the disease. The third herbal extract was an oil obtained from a tree that is indigenous to India as well as China; the Chinese found it effective against leprosy, and even now the antileptol treatment prescribed by the modern physician is essentially purified chaulmoogra oil.

But of all the early Chinese treatments, the most unusual and surprising was the practice of acupuncture. It has been suggested that this might have begun as a semi-magical rite to drive demons from the body, but it was incorporated into Chinese medicine early on, and soon developed into an elaborate, codified system of treatment. The explanations given for its efficacy were legion, although they all centred round the basic desire to balance the Yin and Yang in correct proportions. At all events, whatever the explanation – and it is still as much a mystery today as it has always been – the system seems to work. Could its introduction in the West two millennia ago have revolutionized, or at least revitalized, Western medicine? The answer is probably yes, but since acupuncture is an ancient medical technique that is being introduced into Western medical practice at this very moment, we are in the interesting position of being able actually to observe the effect of the adoption of a lost discovery on our own times.

Aztecs and other early Mexicans that we owe sarsaparilla as a diuretic for kidney ailments. They also knew and used rubber, applying it medically in the form of large plasters for rheumatism and in cases of pleurisy. But, strangely enough, the West did not learn of rubber from Cortez's men but from the French physician and traveller Pierre Barrère, who in 1743 described how in South America objects were made 'from the juice of a tree'. Even then it was still another fifty years before rubber goods were introduced in Europe.

Because of the geographical isolation of the New World, South American medicine was unknown until Cortez's conquest. Partly for the same reason, knowledge of Ancient Chinese medicine was denied to the West until the seventeenth century AD onwards. The Chinese doctors, it seems, were restricted in what could be done to the body from a surgical point of view, and were also limited in the examinations they were able to make, particularly of female patients. These would point out the position of their aches and pains on the body of a doll. As a result, the doctors learned to set great store by the pulse and the colour of the tongue, and the pulse, particularly, became the primary diagnostic centre. Philosophically, the physician

CHAPTER 7

Tomb-Builders and City Planners

ONE of the more surprising things about the ancient world is how long ago very large buildings were made. Huge structures, challenging the skill and technology of the day, were built well before the first millennium BC in virtually every city of the Mediterranean, Indian, Chinese and South American civilizations. Certainly the urge to build big, to honour one's gods and to display one's wealth and prowess by a vast piece of architecture or in a gigantic monument, is as old as man, and had to be satisfied just as soon as he was able to command sufficient resources. But what is so amazing is what was actually achieved.

The first very large buildings were made of plastered mud or clay, but once the technique of casting brick in a mould had been devised, big buildings with elaborations like recessed wall-faces could be constructed. Three temples like this were built as early as 3500 BC at Tepe Gawra in Mesopotamia. The tallest was thirty feet and their lengths varied from sixty to one hundred feet, but they were soon surpassed. At Erec and Eridu some really large temples were put up, as high as eighty feet and more, and containing rows of columns, each some five feet in diameter; but size was not all, for their walls were elaborately decorated with cone mosaics and bas-reliefs. Like subsequent Sumerian temples, these were stepped structures – ziggurats – with stairs leading up to huge raised stone terraces, the lowest of which might be between thirty and forty feet above the ground. The most famous was the one built at Ur by King Ur-nammu about 2000 BC; it had a total height of over eighty feet and covered an area of almost an acre.

In Egypt, large temples and immense tombs like the giant pyramids began to be a feature of the architectural scene sometime about 3000 BC. Of them all, pride of place is certainly taken by the Great Pyramid at Gizeh constructed by King Cheops (Khufu). The biggest building of ancient times, it covered almost fourteen acres and reached a height of 480 feet. How did the Egyptians build an immense structure like this? We do not know all the details even now. It is clear from what remains that they used huge limestone blocks, but there are still problems over the way they bonded them, over how they managed to slide one block across another, and of the way in which the walls and ceiling of the inner chamber are supported.

Obviously, they used a block and tackle to lift the blocks, but even so the precise method of getting objects so large and heavy up hundreds of feet without a tall crane to help them is uncertain, while their technique of supporting the internal blocks is completely unknown. It may well remain a mystery until the pyramid is, if ever, pulled to pieces. For sliding the heavy blocks over each other without so much friction as to make them immovable, the only possible explanation seems to be that the Egyptians must have used a thin bed of viscid mortar and floated the upper block on this. But whatever were the precise methods they used, these are lost; just how the

left The Mayan Pyramid of the Inscriptions at Palenque in Mexico and **below** an artist's impression of Ur-nammu's ziggurat at Ur in Mesopotamia, built more than 4,000 years ago, are both evidence of advanced civil engineering skill in the ancient world. It has been suggested that there was contact between the Old and New worlds in ancient times, but this is still doubted by many anthropologists.

Egyptians achieved their results and the technological discoveries they made to gain their ends are still unknown.

It was not only religious buildings and royal tombs and palaces that were large; sometimes a utilitarian building might be built of great size, and the pharos (lighthouse) at the entrance to the harbour at Alexandria was a case in point. Designed by the engineer Sostratos sometime about 280 BC, it is the first true lighthouse of which there is any record; unfortunately it was destroyed in an earthquake in the fourteenth century, but all the same we have various details about it. (An eighteenth-century reconstruction is shown on page 8.) It was built in stepped form – this was the only way then to construct a really tall building – on a very large base, and it reached a height of at least 265 feet. The light was generated by a fire that burned some sort of resinous wood to give a bright flame, and the light was beamed outwards by large metal mirrors; contemporary accounts say that it was visible for anything up to 35 miles.

At all events, the pharos was famous, and was even named as one of the seven wonders of the ancient world. Naturally enough, it formed the prototype for other lighthouses to guide ships to harbour entrances. But none of the other lighthouses had any lighting system equivalent to the one at Alexandria; fires were lit in them, but there were no mirrors to direct the light. In fact, Sostratos's lighthouse remained unique until lighthouse-building was taken up seriously in the late seventeenth century AD.

But the ancients did not concentrate only on public buildings. They also introduced many refinements in domestic architecture – albeit mainly for the wealthy – and many aids to more hygienic living that subsequent ages forgot. As a result, when the Roman Empire declined, the West stepped backwards and lived more primitively, without those comforts that the Mediterranean peoples had come to expect as part of everyday life and without the sanitation that had become accepted as common practice. In fact the standard dropped so low that we are now tempted to look upon every modern advance in building techniques, or in comfortable living, as something no one has experienced before; yet in so many aspects this is a gross exaggeration. Of course the ancient world had no electricity and no natural gas or oil industry, but they had a highly sophisticated standard of living.

Take, for instance, the question of heating homes. Today central heating is being adopted increasingly, at least in the cold, damp countries of northern Europe and North America; but this is not new, even if the fuel supply methods we adopt are. The Romans were adept at centrally heating their buildings, and anyone having a villa built had a choice of two methods. One was the channelled hypocaust, probably the method most analogous to present-day pipes and radiators. It consisted of an outside furnace from which hot air – not water – passed into a channel or trench that carried it to the centre of the room (or rooms) to be heated. From the centre of the room, other channels conveyed the air to the four walls, and thence up flues in the walls themselves. The second method, known as the pillared hypocaust, was more akin to the underfloor heating of today. Again there was an outside furnace, but instead of feeding channels with hot air, the entire output of heated air was passed into a low basement under the rooms to be heated. The basement, which had a concrete floor, was about two feet deep, and all over it were small pillars some eighteen inches apart. A flagstone floor was laid on top of the pillars and it was this whole floor that was heated. In Germany, Gaul and Britain this system was highly effective and provided a standard of comfort that the medieval and Renaissance times never knew. Only in the nineteenth century did the spartan attitude of the northern Europeans and Americans succumb enough to submit to a rediscovery of the convenience of centrally heated homes.

The Romans' general concern with, and ability to provide, high standards of housing, at least for the well-to-do, was astonishing and far exceeded anything available even to royalty of later ages. For example, it was not until the late 1560s that English houses began to have window glass, and another century before this comfort – or, as we should now say, necessity – penetrated into the colder regions of Scotland. Yet if glass windows were uncommon before the third century BC this soon changed, and by the time of the destruction of Pompeii in AD 79 they were ubiquitous. Most Roman window panes were about twenty-one inches by twenty-eight – quite a large size – and in one bath-house a sheet twenty-eight by forty inches, half-an-inch thick and frosted on one side, has been found. This sheet, and others like it, must have been cast, and probably rolled as well; they would therefore be akin to plate glass today. But most Roman window glass was thinner – only about one-eighth of an inch thick – and was slightly bluish in colour. Much of this also was cast, although some was blown and had a roundel in the centre of the pane. And, as we might expect, Pompeii was no isolated case. Even as far afield as Britain the Roman houses had window glass, although here it was often confined to the windows under the eaves of the outside wall surrounding the peristyle.

Another problem facing the Roman builder, especially when he was building in soggy ground, on land liable to flooding, or even just in general humid conditions, was the perennial one of rising damp. From what Vitruvius tells us, there were a number of ways of preventing this. Provided the damp was not too severe, the walls were rendered with a plaster containing burnt brick instead of sand. But if the damp was excessive, then they built a cavity wall, with vents top and bottom to allow air to circulate, a technique that is once again being used today. Should there not be sufficient space, Vitruvius recommended what is essentially another modern technique, a vertical damp course. The wall was to be fitted all over with flanged tiles treated inside with bitumen: a thoroughly effective method, replaced now by plastic instead of bitumen-coated tiles.

Terraced mountain sides in the Urubamba
Valley of Peru were constructed by the Incas.
As well as building beautiful cities with
magnificent buildings, the Incas united the
Peruvian tribes, enabling them to treat whole
valleys as agricultural units, with extensive
irrigation schemes. As a result, for the
centuries the Inca empire lasted, nobody went
short of food.

In the Roman Empire, cities were built for convenience and comfort, and the wealthy at least, had homes of great refinement. It is, of course, Pompeii that gives us the most complete picture of Roman life at the zenith of the empire, for it was engulfed in ashes in AD 79, when Roman power was supreme.

left The streets of Pompeii were well paved, and had raised pavements (sidewalks) so that traffic and animals were separated from pedestrians.

below The courtyard of the house of a well-to-do citizen of Pompeii shows the high standard of design, well suited to the climate. Glass windows some two feet square were common in Roman cities, yet from the Dark Ages to the sixteenth century window glass was unknown in Britain.

above Throughout the Roman Empire, from the Mediterranean to northern Britain, houses were built with a central heating system – the hypocaust. The type shown at this site in Yugoslavia had a low basement, through which hot air from a furnace passed. In the complete house, a paved floor rested on the pillars shown here.

right A street scene in the Indus Valley city of Mohenjo-daro shows the astonishing standards of public hygiene attained in this civilization over 4,000 years ago. A main drain runs down the side of this major street, with drains from the side-streets running into it. These drains were all covered with bricks, as is shown in the background of the picture.

In many other ways also the Roman architects and builders were far closer to their modern counterparts than the medieval builder ever was. For instance, it was not an uncommon practice to fireproof the timbers to be used in building construction. This was usually done by saturating the wood with alum, a trick that was not confined to the Romans alone. When their army attacked Piraeus (the port of Athens) in 86 BC, they found themselves unable to burn down a wooden tower whose timber had been treated in just this way. Another interesting example of fireproofing techniques is given by Vitruvius in his advice about the construction of heated baths. He recommends making the ceilings of concrete, but, where this cannot be done and a timber ceiling is laid instead, then iron bars or arches must be hung close together underneath the timber so that flat tiles may be carried by them to protect the wood. This method, as one modern authority has remarked, closely resembles some present-day fire-resisting floors.

If the ancient world had some advanced ideas about building, it was also ahead of the civilizations of the first fifteen hundred years AD in the matter of sanitation. At least as far back as 2300 BC the Akkadians built an elaborate sanitary system in the palace of Esh-nunna (now Tell Asmar in Iraq). Lavatories and bathrooms were arranged to lie along an outside wall and discharge their effluent direct into a large vaulted main sewer that ran the length of the street beyond. The bathrooms had floors and surrounds of bitumen-covered brick, and the lavatories all had raised stone pedestals, some with shaped seats made of bitumen. As far as the drains were concerned, these were of baked brick, joined and lined with bitumen, with inspection chambers at the principal junctions. In addition there were inspection 'eyes' that allowed drain-rods to be inserted for cleaning. In fact, the whole system was a pre-echo of modern sanitary engineering. It even exceeded the standards of the Assyrian palace of Prince Zimrlim of Mari built 600 years later, where there were ceramic bath tubs, but drains that were not as elaborate – though they were large enough for young slaves to crawl through to clean them.

Certainly, most of this early elaborate sanitation was for the few – for royal palaces and the homes of the nobility – but there were exceptions, exceptions that should have been followed by later generations but which were neglected so far as the West was concerned. In the Indus Valley civilization, at Mohenjo-daro, in the third millen-

right A Greek vase painting of the sixth century BC shows women taking a shower. The Greeks realized that pure water – for both drinking and washing – was essential to health, and they built efficient public water systems. By using syphons (a Greek invention) they avoided excessive tunnelling or building aqueducts via long, roundabout routes.

below The Pont du Gard, near Nîmes in southern France, is one of the best known surviving Roman aqueducts. The methods of supplying water to cities originated in Mesopotamia, but spread most widely under the Roman Empire. Less spectacular than the soaring aqueducts, but equally great in civil engineering terms, were the miles of underground water pipelines.

nium BC, there was a public sanitation system similar to the kind we should provide in a city today, and surpassing anything in Babylonia or Egypt. There were plenty of sewers, made of brick with a bitumen finish, and these drained not only the main streets but the side streets as well. They were big enough for a man to walk through them standing upright. From the houses, ceramic drains ran to the sewers. There were underground water pipes, too. The city also boasted large public baths, with changing rooms, fountains and even steam baths, while there was also a huge swimming pool with pipes and drains for changing the water. The quality of construction may be judged from the fact that the swimming pool is still watertight, 5,000 years after it was built. But perhaps ancient sanitation reached its peak in Rome, where, by AD 315, there were 144 public lavatories flushed by the public water supply.

The contrast between Rome and a medieval city is marked; between Mohenjo-daro and, say, medieval London it is astounding. The retrogression from the ceramic drains and brick sewers of the Indus Valley city, and the slops tipped out into open sewerage channels running down the streets is utterly astonishing; if it were not a matter of known fact it would be unbelievable. Yet if there was a deterioration in sanitation conditions, this was paralleled, even if not quite so severely, in the matter of water supply.

Many ancient cities had supplies laid on, in some cases from a considerable distance, but of course they varied in size and elaboration. One of the most remarkable arrangements ever built was the stone canal that King Sennacherib constructed in the seventh century BC. Connecting Nineveh with a source of supply some fifty miles away in Bavia in the northern foothills, it mainly consisted of a channel lined with more than two million stone blocks, each of some six cubic feet. At one point it crossed a river 300 yards wide, and the channel itself was in many places twenty yards or more across. There were dams and sluice gates to regulate the water flow, while an additional dam outside the city walls deflected some of the incoming supply to irrigate orchards and gardens. And this aqueduct and artificial canal were the more remarkable in that the whole scheme was built in no more than fifteen months. It may have been built with slave labour – and probably was – but the organization needed to plan, and then carry through, such a programme is breathtaking. It is on a par with the planning that must have gone into the building of the Great Pyramid at Gizeh.

The Nineveh water supply was not unique. Other Assyrian kings built similar, if not such extensive, systems, and Sennacherib's engineers furnished the city of Erbil with a supply from the Khani mountains, a dozen miles away, because the water from the Tigris river was unfit to drink. And the Greeks, in spite of the distaste their philosophers had for practical matters, were not at all incompetent when it came to supplying water to their cities; in fact they applied their invention of the siphon to obviate the need for so much tunnelling, or the long winding routes that so often had to be taken by a system based solely on a gravity flow. For example, the city of Pergamum, which was built on a hill some 300 feet above the main supply line, was by 180 BC provided with a good siphonic supply travelling through metal pipes. What metal we cannot now tell, since none of the actual piping remains. Could it have been copper? Or perhaps lead? We cannot be sure, but we do know that the Romans later found lead unsuitable for piping drinking water since it caused lead poisoning!

The Romans, of course, are remembered for their efficient and elegant aqueducts – although it is the beautiful arrangements of arches that have attracted later artists, not the miles of underground piping that was also needed. But beautiful or hidden, the system they devised for Rome was applied throughout the Empire – in Britain, in Gaul, in Germany and elsewhere. To gain some idea of the scope of their waterworks, and the administration involved, it is worth noting that in Rome, by the fourth century AD, the Emperor was allotted seventeen per cent of the supply, private houses and industrial concerns made up a further thirty-eight and a half per cent, while public supplies accounted for the rest (a little over forty-four per cent). Among the public uses were supplies to ninety-five official buildings, and no less than 591 cisterns and fountains. Yet in medieval times the Roman system, in those cities where it existed, was allowed to fall into disrepair. There was no central organization to control the supply or supervise the upkeep, as there had been with the Romans, and this is yet another aspect of the way in which the standard of living in medieval times fell below that of the ancient world. In the matter of water supply it is perhaps epitomized by the fact that the medieval archbishops of Salzburg had to have their supply of fresh water brought daily by special carrier.

What strikes one about ancient cities is the sound and efficient way they were planned. In one sense they started like any medieval city, with a temple the focal point, and in the ancient city probably with a large granary for storing the supplies surplus to immediate needs. But the difference arose because, whereas the medieval city seemed to grow in a haphazard fashion around its cathedral or its palace, the ancient city had more often than not a planned development. Babylon, for instance, had a central area occupied by a giant ziggurat, and around this was a network of streets all at right-angles to one another; they were built in fact on a grid system, so beloved of modern American city planners. The city of Lagash was similar, while the Egyptians and the Chinese adopted the same kind of basic design. But probably the earliest city to have a rectangular grid of streets crossing and re-crossing one another, the archetype of all later cities laid out in this way – even of New York – was Mohenjo-daro. This had one broad main street and, parallel with it, some other broad but secondary thoroughfares. The rest of the streets were neatly placed at right-angles.

An artist's impression of the palace of the
Assyrian kings Tiglath-Pileser III and
Ashur-nasir-pal II at Nimrud shows the
exquisite architecture of the great buildings of
the ancient world. Sited beside the River
Tigris, the palace was surrounded by
beautiful gardens where decorative and
medicinal plants could grow, even amidst an
arid land.

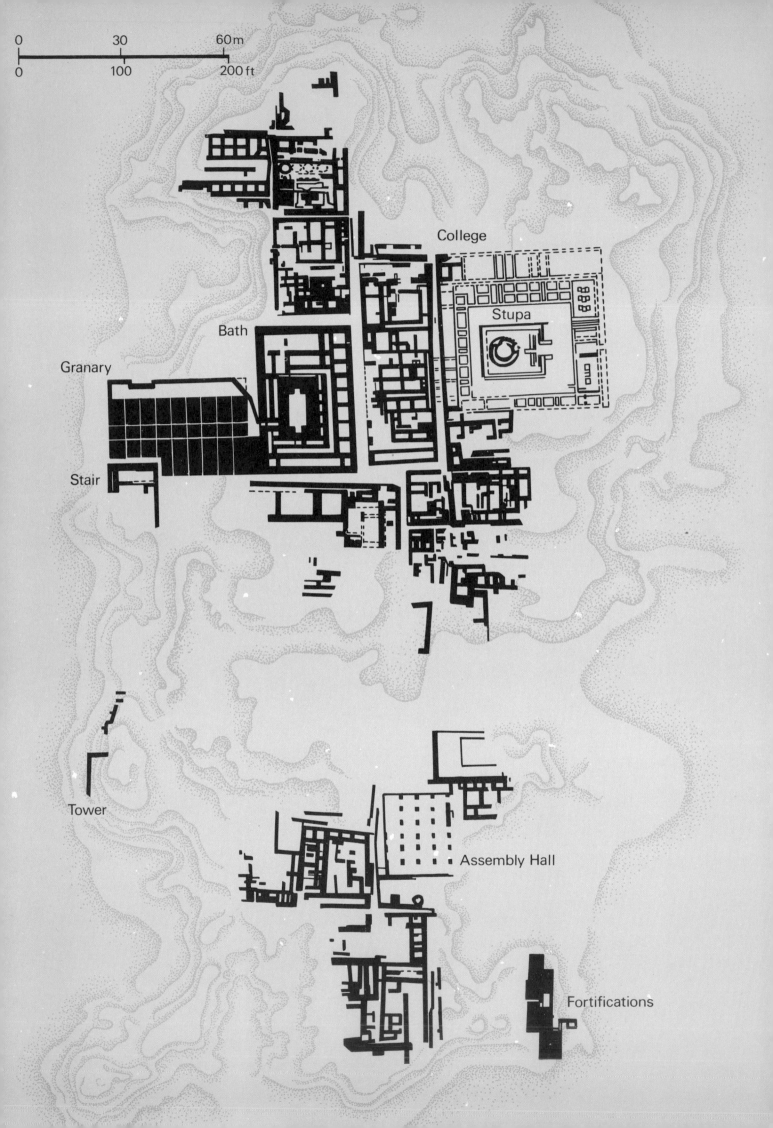

0   30   60 m
0   100   200 ft

College

Stupa

Bath

Granary

Stair

Tower

Assembly Hall

Fortifications

left A plan of the Indus Valley city of
Mohenjo-daro, based on the work of Sir
Mortimer Wheeler : The city was built on a
grid plan very much like that of many modern
cities, unlike medieval towns which grew up
piecemeal.

Looking back on these examples of town planning – and not only on the grid pattern of streets, but also the provision of sewers and fresh water – one is struck by the immense organizational ability that was not only available, but was applied to practical everyday tasks. So often the ancient world is presented to us as a source of art, of letters, of philosophy and of science that, by their magnificence and their scope, can leave us breathless. But it is good to remember, too, the armies of administrators and craftsmen that saw to it that the other, more prosaic departments of life were also conducted equally smoothly. And it is also easy for us today to neglect the vast amount of intercourse there was between one city and another, and the comparatively high degree of travel and trade between one country and the next. Although many trade routes were by sea, and some lay across wasteland or desert, it is surprising how many roads were actually built and how extensive some of them were.

Once the wheeled vehicle replaced the sledge in about 3000 BC, there was a need to transform the rough, unmade tracks into properly constructed roads. But since so much inland traffic was carried by water, the urge to build good roads for commercial purposes was not strong. It was the Assyrians who first began to organize a road-building programme in earnest, and then it was for military reasons. In the first half of the first millennium BC they began to level tracks and compact their surfaces, setting up watch-towers along each route at regular intervals. This example was followed by the Persians who, in the fifth century BC, with a great empire to administer, needed inland routes for trade and administration, as well as for military use. Rest houses with guards were established every fifteen miles, and there was a continuous unmade but levelled road that ran for 1,600 miles all the way from Sardis, on the east coast of the Mediterranean, to Susa. A special messenger, it was said, could cover a distance of 100 miles a day, the road was so good. All the same, this great route was really no more than a levelled track, and when the Persian empire collapsed the road was still unpaved. It fell into disuse, to become overgrown and lost.

Meanwhile, in China, a vast road-building programme had been in operation from about the ninth century BC. Roads were classified

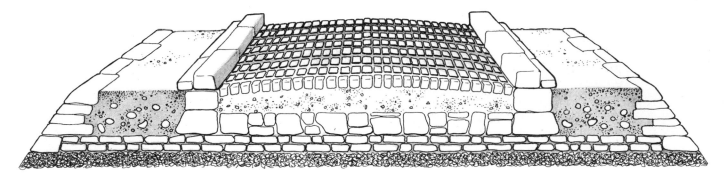

Roman methods of road construction were highly sophisticated. The first diagram shows a major highway in Italy. It is metalled with cobbles or stone slabs set in mortar. These rest on layers of concrete, stone blocks set in mortar and, at the bottom, flat stones. Curb-stones bound the sides of the roadway. The second drawing is of a normal Roman highway with either a gravel (left) or a stone-sett surface. This again rests on a layer of concrete, followed by stone slabs in mortar and a bottom layer of sand.

into five categories, each of a standard size; some were mere pathways for men and pack animals, others roads for narrow gauge vehicles, then roads for one, two, or three lanes of normal traffic. Moreover, the Chinese appointed traffic commissioners whose job it was to see that vehicles kept to certain prescribed standard sizes, and to prevent furious driving. The broadest roads were fifty feet wide, and trees were planted along them every thirty feet. The Chinese road-building programme took time, but by AD 190 a total of 22,000 miles had been constructed.

As surprising as the extent of the Chinese roads was the way they were built. The road bed was levelled and filled with rubble that was then tamped down with metal rammers. After this a special gravel mixture was laid and again tamped down. When finished, this gave a slightly pliable surface, what one historian of technology has aptly termed a 'flexible water-bound macadam'. And it is no exaggeration to say that it was in China that the best roads of antiquity were laid, roads that were not bettered until the macadamised surface was invented in the late eighteenth century AD.

Roman roads were, of course, of high quality, and perhaps close to the Chinese. Some of them were paved, but in general they adopted a method of construction not dissimilar to the Chinese, especially for their gravelled roads which formed the major part of their road system except in Africa and Syria where the predominantly dry conditions made unmetalled roads practicable. But in spite of the extensive road system in the Roman Empire, and the vast network in ancient China, one of the most surprising – perhaps the most wonderful – of all the ancient roads was the Royal Road of King Asoka, built in the third century BC. It ran from the Indus in the west, across the Punjab, over to the river Ganges in the east, a distance of 850 miles. And what was so remarkable about this road was not its method of construction – it had no water-bound macadamised surface like the Chinese built – nor its extent, but the way it was laid out with wells, rest houses, shady places and gardens of herbs for the health of man and beast as they journeyed along it. When we consider the way in which the West failed so dismally to preserve what the Romans left, and the hazards of cross-country travel, it is perhaps salutary to turn back to Asoka's own words, written over 2,000 years ago, words that embody his attitude to the well-being of his subjects at home, and as they travelled along his royal road:

Everywhere . . . has king Asoka . . . established medical treatment of two kinds, medical treatment for men and medical treatment for animals. Wherever medicinal herbs, wholesome for men and wholesome for animals, are not found, they have everywhere been caused to be imported and planted. Roots and fruits, wherever they are not found, have been caused to be imported and planted. On the roads, wells have been dug and trees caused to be planted for the enjoyment of man and beast.

opposite This 40-foot-high stone column was erected at Laurya Nandangarh in Nepal in 243 BC. It is one of many such pillars built by King Asoka, all carrying Buddhist inscriptions. Asoka established a great 800-mile-long road across India, providing at regular intervals places where men and animals could rest in shady gardens where medicinal herbs grew.

Many of the major mechanical developments
of the ancient world were in the field of war
machines. This Assyrian relief shows one of the
earliest : a siege engine pouring boiling oil on
the enemy during Sennacherib's siege of the
city of Lachish in about 700 BC.

CHAPTER 8

Machinery in the Ancient World

COMPARED with so many creatures, man is a puny being with very limited physical powers. But as far back as we can trace him he has compensated for his lack of brawn by applying his ingenuity; he has always used the natural world for his own protection, for his profit, for his comfort. This is how he has survived. His use of tools and of fire, his taming of wind and water, have all lifted him from the status of an animal to a stage where he can come to terms with his environment. His motivation has always been practical and technical developments have usually been prompted by the needs of the day. When a large slave force was available there was little point in devising ways to lighten manual labour, and the one fundamental reason for machinery was to tackle those tasks that even large, organized teams of men could not accomplish. This is how it was used in Egypt and Babylonia – as a utilitarian adjunct to man, as an aid to building, to raising water, to waging war. Only with the advent of the Greeks did other uses and different attitudes arise.

The Greek civilization was to a certain extent slave-oriented, at least from the fifth century BC onwards. Since many (although not all) manual tasks were done either by slaves or hired labourers, an attitude grew up that manual work was lowly, or at least of a lower kind than that of the thinker, the philosopher, the man who did not dirty his hands with physical tasks, but concerned himself with the superior activity of the mind. And this attitude was not confined to Greece, nor to the Mediterranean; the superiority of a special class, philosophical or just leisured, was ubiquitous, although in Greek hands it became embodied in a whole cultural outlook. Yet there were always some philosophers who did not run true to type, who did design mechanical devices, who did, perhaps, gain some satisfaction from seeing a piece of machinery in operation. And there is evidence such as that of the computer salvaged from the Antikythera wreck (Chapter 4) which may mean that there was a technological sub-culture of which we know little. At all events, it is in Greek times that we find a development of machinery and the beginnings of true mechanization.

The Greeks took over a heritage of very simple machinery, of devices that were unsophisticated, even though they were of basic significance. Probably the most important was the lever, discovered as soon as man had learned to dig the ground with anything besides his hands, and used extensively in Egypt and Babylonia. So, too, the Greeks knew about the wedge, the first cousin to the lever, again with its origins in prehistory, and the wheel that as far as one can tell was a Sumerian invention way back in the fourth millennium BC. The pulley wheel was another important inheritance, this time probably from Assyria – although it, too, might have been Sumerian, or perhaps Akkadian. The last basic machine the Greeks inherited was the so-called bow-drill. It was used for everything that needed holes boring in it, in making furniture, chariots, and the hundred and one tasks that we use a drill for today. And the same principle, utilizing the reciprocating motion of a bow and string to make a spindle rotate, was applied to the lathe. Early on, this could only be used for woodwork and was not suitable for making metal screws or gears, which are a development of the screw. This had to be done by a quite different method, and it may not have been until the fourth century BC that the Greeks devised a way of doing it.

In view of its importance, and the machinery that developed from it, it is worth looking at the problem to be overcome and the ingenious solution that was found. The difficulty was to trace the correct spiral, the precise line, around and along a rod to act as an accurate guide line for the actual cutting process; if this were not accurate then no amount of subsequent cutting, however careful, could make it so. What the Greeks did was to take a sheet of soft metal (probably copper) and cut it to the exact shape of a right-angled triangle. They then laid one of the smaller sides along the length of the rod and, once this was carefully positioned, wrapped the rest of the metal sheet round the rod. The hypotenuse (the longest side) then traced out a spiral round the rod, and this could therefore be accurately marked and then painstakingly cut. And once an accurate screw had been cut, this could be used for marking out gear teeth for the subsequent cutting of precision gears. This was an elegant solution to the problem, and one that displayed the Greek flair for geometry to perfection. But who invented this beautiful method we do not know for certain, although it is often attributed to Archytas, a Pythagorean philosopher and mathematician who worked in Tarentum sometime around 375 BC.

Without doubt, one of the great Greek inventors was Archimedes, theoretical mechanician though he might have preferred to be called. There has been doubt cast on whether he was the Father of his best-known invention, the Archimedean screw, which was used for raising water; it was found everywhere in Egypt in the third century BC, and it has been suggested that it might have been attributed to Archimedes. However the historian Diodorus Siculus, writing in the first century BC, claims that Archimedes did invent the device, but while he was on a visit to Egypt; and modern research has tended to confirm this. But whatever the final verdict about the screw, there is no doubt that it is to Archimedes that credit must go for the compound pulley. Plutarch has a nice story about this. He says that King Hieron asked Archimedes to show him how great a weight could be moved by a small force, and Archimedes picked on a three-masted merchant ship of the royal fleet that had been dragged ashore after immense effort by a large labour force. He had the ship loaded with passengers and its usual amount of freight, and then sat down some distance from her. With no apparent effort, he quietly began to set in motion a system of compound pulleys, and pulled the ship smoothly towards him 'as though she were gliding through the water'.

The compound pulley is a collection of pulley wheels arranged in

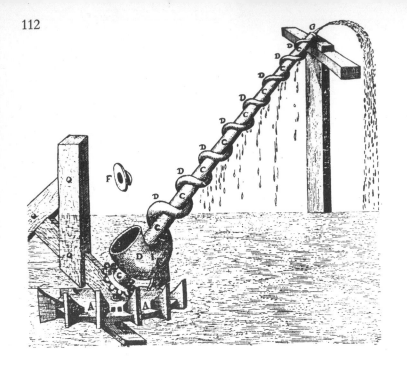

Three great names stand out in the field of mechanics in Ancient Greece: Archimedes, Ktesibios and Hero.
left Archimedes is perhaps best known for the Archimedian screw, although there is some doubt, in fact, as to whether he did invent the device. At all events, it was still in wide use after 2,000 years, when this early seventeenth-century illustration was produced.
below A sixteenth-century reconstruction shows Ktesibios in his room behind his father's barber-shop. On the wall just to the left of the window is shown the adjustable mirror with pneumatic damping that Ktebisios made for his father. This led to the development of pumps for both air and water.

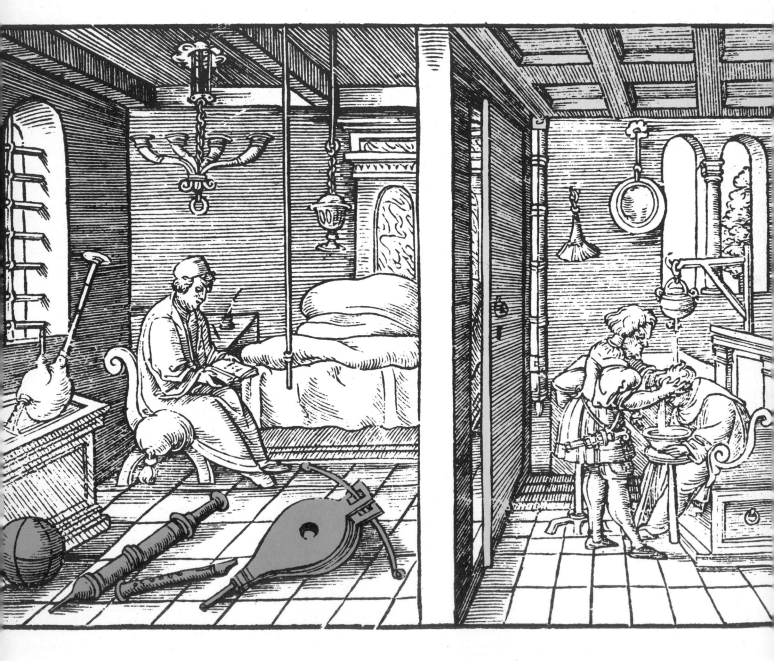

two blocks, so that a rope passes in over one pulley wheel, down to one below, up again to a second in the upper block, then down again, and so on. The fact that Archimedes had appreciated – and which, of course, he was to deal with theoretically – was that the greater the number of pulley wheels, the smaller the effort required to move a given load. For example, if the upper pulley block has two pulley wheels and the lower block two, then only a quarter of the effort is needed to lift a weight compared with the effort if there were no pulleys at all. Should there be three pulley wheels in each block, then the effort required would only be one-sixth. Plutarch's story may be an exaggeration, but what he said is sound enough in principle, and quite possibly Archimedes was not averse to presenting the results of his research in a dramatic way. The compound pulley became widely used after Archimedes, especially in cranes for shipbuilding and other constructional work. These cranes were sometimes worked by a straightforward winch, sometimes by a treadmill.

The second great Greek mechanician of this period was Ktesibios, a slightly older contemporary of Archimedes and the builder of many water-clocks, as described in Chapter 2. To him we owe some fundamental work on air and water machinery. The son of a barber, he is supposed to have come upon his discovery that air could be compressed when he made an adjustable mirror for his father's shop. The mirror had a counterpoise which consisted of a lead ball moving up and down inside a tube; as it moved the ball compressed the air, which escaped with a loud noise. This led Ktesibios to devise the cylinder and plunger, and so to force and suction pumps for both air and water. These became universally used for raising water in mines, for irrigation and for drinking. In short, they were one of those basic devices, like the pulley, that lie behind a vast amount of the subsequent mechanization that was developed in the Western world.

But one fact we must face is that much of the early machinery had no utilitarian side to it at all. It was made to amuse, to intrigue, to impress. Much of the mechanician's time was spent designing devices that would fascinate, not machines to do a job of work. And if such an attitude seems rather foreign to us now in our automated world, it certainly should not surprise us; today we make use of all kinds of machines for our entertainment, from automobiles and motorcycles to television sets and tape recorders, and the ancient world was merely doing the same. The devices were different, and only reached the homes of the well-to-do or some selected public buildings, but the principle was identical.

How much or how little Ktesibios was concerned in devices of this kind we do not know, since his original writings are lost, although there is evidence that he made a forerunner of the mechanically pumped organ. But the main source of mechanical devices in ancient times is Hero of Alexandria, who lived and worked there during the first century AD, lecturing on mathematics, physics, pneumatics and mechanics. Some of the devices he describes are probably due to predecessors like Ktesibios, but this does not matter since we are more concerned with what was actually done than with who was responsible.

Hero did design some purely utilitarian machinery, a task that every Greek mechanician seems to have taken on whenever a real demand arose. Among these devices was one for cutting screw threads in wood, a syringe, a fire engine containing a pump after Ktesibios's design, the hodometer (taximeter) – perhaps the one the Romans used (Chapter 4) – and an oil lamp in which the wick was trimmed automatically. He also designed a successful surveying instrument, the dioptra; this contained some very accurately cut gears, and seems to have been something like a cross between the surveyor's level and the theodolite of later times. Hero was also responsible for an efficient piece of war machinery that we shall come to later, but at the moment we must look at his automata and other inventions of a less serious kind.

The fascination of automata never seems to have lost its hold, and I myself still recollect how impressed I was when, as a child, I saw at an exhibition four old ladies, each about six inches high, seated round a table pouring out tea and then drinking it. In days when any mechanical device was a marvel, how much more astonishing Hero's 'toys' must have seemed. Indeed, they so intrigued his contemporaries that they too write of them and, when the Renaissance came, one author at least was so carried away that he illustrated and described them as examples of ancient ingenuity and, so, of ancient wisdom. Naturally enough they varied widely in complexity. There were birds that sang as soon as water was poured into a container; others that moved about as well as singing; animals that drank when presented with water; satyrs that poured water from wine-skins into basins that mysteriously never overflowed; a little altar on which, once a fire had been lit, a serpent hissed and human figures poured out libations; and, perhaps most ingeniously of all, a model in which once an apple was picked off a tree, a figure of Hercules struck down a dragon.

Hero also built two automatic puppet theatres, both driven by a weight that rested on a heap of millet seeds that escaped through a small hole; the weight was attached to a rope wound round an axle, and it was the axle's rotation which caused the whole sequence of subsequent events. One threatre moved along a track, displaying an altar with a fire on it, and Dionysius pouring out a libation, while bacchantes danced around him to the sound of trumpets and drums. The other was stationary but more elaborate. Doors opened and closed automatically at the beginning and end of each performance, which depicted the well-known story of the wrecker Nauplius, son of the sea-god Poseidon, and the warrior Ajax. The sequence of actions showed shipwrights at work, the launching of a ship and its subsequent wreck due to a false beacon lit by Nauplius, and ended with thunder and lightning and the appearance of the goddess

Athena, who then destroyed the warrior Ajax who had been on board the ship. The ingenuity of the mechanisms of levers and pulleys by which these theatres operated – there were no gears – was astonishing, but the secrets were forgotten by the West until the thirteenth century, even though Greek manuscripts describing the devices survived the destruction of Alexandria.

The appearance of Athena towards the end of the Nauplius performance echoed a tradition in the (full-sized) Greek theatre not only that the gods should be seen performing their divine acts, but also that they should appear, appropriately enough, out of the sky. For this the Greeks devised a kind of crane with a long jib named a 'mechanee' – which means, literally, a crane's bill – coupled to a lifting device with pulleys. They also had sliding scenery used for uncovering something previously hidden behind it; a sliding wall of a house to disclose a room inside, for example. The producers of the Greek theatre became adept at using stage machinery of a really elaborate kind, yet this was passed over and forgotten in the West until well after the dawn of the eighteenth century.

Many of Hero's devices were aimed at performing what seemed to be miracles, and demonstrate that there was clearly a side of temple worship in which the intention was to impress and amaze the faithful. In days when the most sophisticated piece of domestic apparatus would be a hand-mill for grinding corn, even a comparatively simple thing like a penny-in-the-slot machine that dispensed lustral water

was considered astonishing, and in his *Pneumatics* Hero describes one of these in detail. But perhaps the greatest of his 'miraculous' machines was the underground system of pulleys and pipes that operated temple doors so that these swung open when fire was lit on an altar, and closed when the fire went out. The method was basically simple, but no less ingenious for that: When the fire was lit the heated air inside the altar expanded and forced water out of an underground spherical tank into a bucket; this pulled on ropes and opened the doors. When the fire went out, the air contracted, and water was sucked back from the bucket through a siphon and passed into the sphere again; the bucket moved upwards and so made the ropes close the doors. This was not a full-scale arrangement but a miniature pair of doors placed, with their altar, on a pedestal that contained the secret mechanism, but it was all the more effective since it was clearly too small to hide a human being and must work by some form of 'natural' magic. But devices like this were eschewed by the Western Christians, and not until the sixteenth century do we find them again.

Machinery was used for the art of war comparatively early in ancient times. The Assyrians in particular devised wheeled towers and what, for want of a better term, we can call wheeled tanks. Both

right Among the best-known of Hero's inventions was this pneumatic organ. It was operated by keys, whose mechanism is shown on the left. The air was pumped by hand, using the lever-and-plunger arrangement illustrated beneath the array of organ pipes. This is a nineteenth-century reconstruction.

opposite An artist's reconstruction shows Hero's model temple with automatically opening doors. All that needed to be done was to light a small fire on the altar; operating by the expansion of heated air, the doors were 'magically' pulled open, only to shut again as soon as the fire went out.

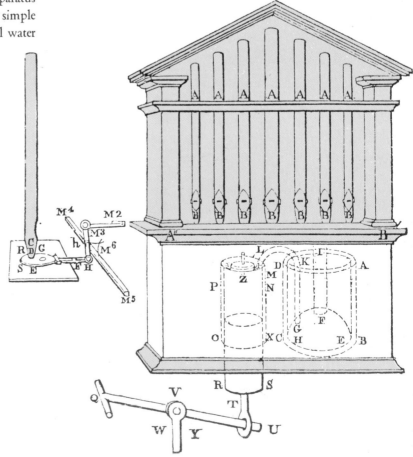

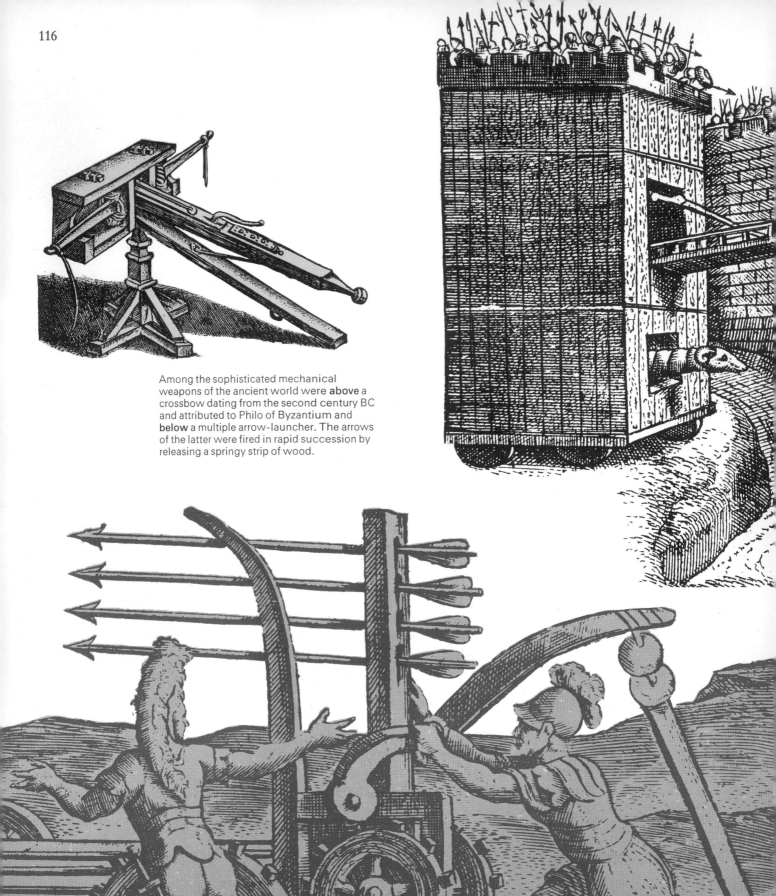

Among the sophisticated mechanical
weapons of the ancient world were **above** a
crossbow dating from the second century BC
and attributed to Philo of Byzantium and
below a multiple arrow-launcher. The arrows
of the latter were fired in rapid succession by
releasing a springy strip of wood.

The Romans (who, like the Assyrians, had
highly-organized armies) perfected many war
engines originated by the Assyrians.
left This wheeled siege tower had a battering-
ram for attacking city walls, while troops could
storm the defences from the high platform.

were primarily used for assaulting defences, the first carrying soldiers
who could be brought level with defended walls so that they could
storm in over the top, the other carrying armed men who could use
the tank as a battering ram. A third war engine was a covered device
for undermining the walls of a fortified city. This had a roof to shield
the engineers beneath, who could mine under the city walls, under-
pinning with wooden props as they went. When all was ready, the
props were set on fire, a section of the wall collapsed, and troops
could storm the gap in the general confusion. But these assault
vehicles, which were made great use of by the Greeks and were
developed still further by the Romans, were not the only mechanized
devices that the ancient commanders could call upon; they had
mechanized weapons as well. These were essentially of two main
kinds, catapults and arrow launchers.

The mechanized catapult was the artillery of the ancient world.
Often it consisted of some mechanism for drawing back a springy
wooden arm that could be wound down into position so that, when
it was released, it would eject a heavy stone, pebbles, iron bolts or
other convenient missiles at the enemy. But in some cases the device
was driven by a weight rather than a spring. Here the missile was
placed on one end of a pivoted arm which carried the weight at the
other. The weight was hauled upwards and held in its elevated
position. The missile – perhaps a barrel of 'Greek fire' (whose main
constituent was naphtha) for setting fire to enemy fortifications – was
then loaded and the weight released, when it would swing downwards and send the missile on its destructive course. These 'ballistae'
were basically simple devices, but they enshrined sound mechanical
principles, and were continually being improved by men like
Ktesibios, Hero and Philo of Byzantium, who lived and worked in
the latter half of the second century BC.

The arrow launchers tended to be rather more complex than the
ballistae. Most were small and portable, being really various forms of
crossbow. One of the most efficient of this kind was the 'gastraphetes'
that was, as its name implies, fired from the stomach. Designed originally in the first century BC by Zopyrus of Tarentum, it was
improved by Hero, who used a toothed rack and pawl for holding
the bow-string in tension. Larger, stationary cross-bows used a small
winch to wind up the bow-string, and Philo even went so far as to
design an automatic bow in which a wooden spring replaced the
bow-string and arrows were automatically fed into the loading
chamber each time the spring was drawn back. And at some stage
this was developed into a winch-wound multiple arrow launcher
that could shoot four very large and heavy arrows at once. Yet, like
the siege machinery, these mechanical weapon developments were
not followed up after the fall of Rome and the disintegration of the
Roman army. Thus, for example, the longbow was the most sophisticated weapon used in the Battle of Hastings, and only in the thirteenth century AD did machinery again begin to be applied to war.

below Hero also developed the very first
steam engine, the aeolipile, which worked by
jet propulsion. This was, however, more of a
toy than a practical machine. It was very
inefficient, there was little need for a power
source of this kind in the ancient world, and it
was, in fact, never put to use.

Does this represent the beginnings of mass-production? An artist's reconstruction shows the great flour mills built by the Romans at Barbegal, near Arles in southern France, early in the fourth century AD. It consisted of seven mills, interlinked by internal stairways. The millstones were driven by wooden undershot water-wheels (show in detail **far right**).

lower right The turning movement of the water-wheels was transmitted to the huge millstones by iron shafts and wooden cog-wheels. The total output of the mills could reach several tons per day – far more than enough to supply the town of Arles. Presumably, this was a distribution centre for a far wider area, and may also have supplied units of the Roman army.

Machinery might bring great advantage in war, but in peace it was generally neglected. Wind-driven and animal-powered machinery was familiar, but although Hero devised a simple and effective steam jet machine – the aeolipile – that demonstrated the power of escaping steam to do mechanical work, this was never taken up as a serious source of power either in his own time or later – at least until the advent of steam power in the late seventeenth century AD. The most powerful energy source in the ancient world was water, and this was utilized by making it drive various kinds of water-wheel. An early type was the 'Greek mill', in which the blades were curved and the axle of the wheel was vertical. It required running water, but not a regulated supply as is the case with other types of water-wheel, and was essentially the forerunner of the water turbine. But the most common type that was later to be used so widely in Greek and Roman times, and in the Middle Ages too, was the vertical water-

A seventeenth-century illustration shows the most common source of mechanical power in Ancient Greece and Rome: the undershot water-wheel. Here it is being used in a forge to operate two pairs of bellows.

wheel. Here the water could drop down on the blades from above – the 'overshot' wheel – or sweep in from underneath, giving the 'undershot' wheel. The water-wheel became the ubiquitous method for grinding corn among the Greeks and Romans, but in late Roman times there was a further development in parts of the Empire that seems astonishing to us today, and this was the multiple mill that gave rise to the first mass-production factories.

In the south-east of France the Romans established a number of flour-grinding mills. At Barbegal, close to Arles, there was a sloping conduit that came off the Les Baux aqueduct, built early in the second century AD. About AD 310 this channel was used to feed water to drive a series of seven undershot water-wheels, which, in turn, operated a chain of wooden cog-wheels that drove a series of mill-stones. Very heavy millstones could be used and the output from the factory was high: over a quarter of a ton of flour an hour or 2·8 tons each ten-hour day, an amount sufficient for the needs of a population of 80,000. Since at this time Arles had only 10,000 inhabitants, it is obvious that the Barbegal mill was a production centre for a much wider distribution, and a source of supply for the army, especially since we know that Arles had large complex of grain warehouses.

It is to the Romans, then, that we owe the beginnings of mass production. How far they would have pursued this idea if the Empire had not fallen we can only conjecture. We know that not only was Barbegal no isolated case, but that they were considering the application of multiple water-wheels to other machinery besides that for grinding corn. But in a sense the question is an idle one, for when the Empire finally crashed in ruins early in the fifth century AD, the whole idea fell into oblivion; without a well-organized and extensive state, the demand for quantity production evaporated. Men turned to struggle piecemeal with the problems of living, and the arts developed by the ancient world – in the field of machines as in so many others – were forgotten. Not until the adoption of steam power and its application to all kinds of machinery in the eighteenth and nineteenth centuries were the factory methods of the Romans revived and developed; only then was our modern industrial civilization born out of the ashes of the past.

Two examples of horizontal water-wheels, here used for grinding grain. The wheel on the left is the type known as the Greek mill. It was one of the earliest forms of water-wheel, yet closely resembled modern turbines.

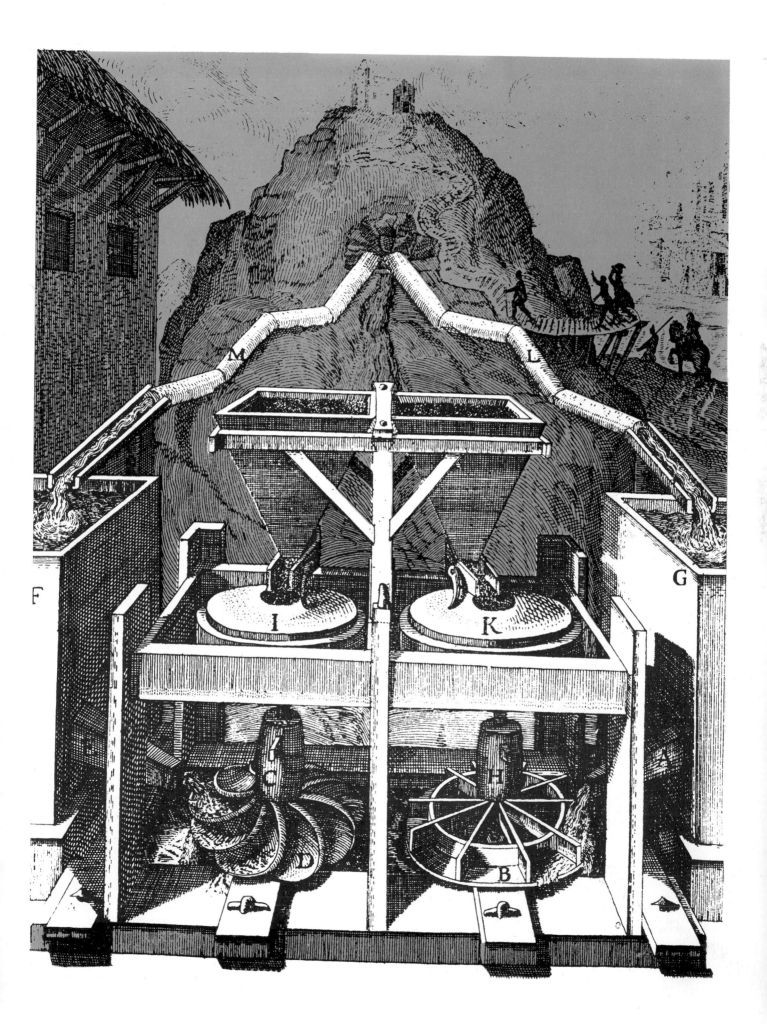

Bibliography

Much of the material contained in this book has been culled from papers in specialist journals and similar sources, and it would be tedious to detail these. But if the reader wants to follow up any subject in more detail, the best approach would be to consult certain reference works and obtain further references from these. For a general look at ancient science, including medicine, it would be best to look at *A History of Science* by G. Sarton, published by Harvard University Press and Oxford University Press: Volume I, 1953; Volume II, 1959. The main reference source for ancient technology is the *Oxford History of Technology*, edited by C. Singer, E. J. Holmyard, A. R. Hall and R. Williams, published by Oxford University Press: Volumes I and II, 1958.

For an account of north-west European megalithic structures, see *Megalithic Sites in Britain* by A. Thom, published by Oxford University Press, 1967; and *Megalithic Lunar Observatories* by the same author, published by Oxford University Press, 1971. For writing numbers and the derivation of mathematical symbolism, there is Karl Meninger's fascinating *Number Words and Number Symbols*, published by M.I. T. Press, Cambridge, Massachusetts, and London, 1969. Lastly, the sole comprehensive work on Chinese science and technology is J. Needham's *Science and Civilization in China,* currently being published by the Cambridge University Press; six volumes have so far appeared.

Index

ACKNOWLEDGEMENTS

The Editors gratefully acknowledge the courtesy of the following photographers, artists, publishers, institutions, agencies and corporations for the illustrations in this volume.

p. 8 Ronan Picture Library
p. 9 Geoffrey Watkinson
p. 11 Professor Owen Gingerich
p. 13 By courtesy of the Trustees of the British Museum
p. 14 Ronan Picture Library
Ronan Picture Library
p. 15 Roger Wood
p. 17 Ronan Picture Library
p. 18 Ronan Picture Library
p. 19 Ronan Picture Library
p. 20 Ronan Picture Library
p. 22 Ronan Picture Library
p. 23 Ronan Picture Library
p. 24 Ronan Picture Library
p. 26 Ferdinand Anton
C. A. Burland
p. 27 By courtesy of the Science Museum/John Smith
p. 28 Royal Astronomical Society
Barnaby's Picture Library
p. 30 Christopher Marshall
Christopher Marshall
p. 30/31 Crown Copyright reserved: Department of the Environment
p. 31 Aerofilms Ltd.
p. 33 Ronan Picture Library
p. 34/35 John Smith
p. 36 Eugene Fleury
Eugene Fleury
p. 37 Eugene Fleury
p. 38/39 Michael Holford Library
p. 40 Eugene Fleury
p. 41 Geoffrey Watkinson
p. 42/43 Michael Holford Library
p. 44/45 Ronan Picture Library
p. 46 Michael Holford Library
p. 46/47 Roxby Press Ltd.
p. 47 By courtesy of the Trustees of the British Museum
p. 48 Staatliche Museen Berlin
p. 49 Ronan Picture Library
p. 50 Ronan Picture Library
p. 51 Christopher Marshall
John Smith
p. 52 Ronan Picture Library
Ronan Picture Library
p. 53 Ronan Picture Library
p. 54 Dr H. J. J. Winter
p. 56 By courtesy of the Trustees of the British Museum
p. 57 Ronan Picture Library
p. 58/59 Michael Holford Library
p. 59 Ronan Picture Library/Royal Astronomical Society
p. 60 Ronan Picture Library
p. 60/61 Ronan Picture Library

p. 61 Ronan Picture Library
p. 62 Robin Dodd
Spectrum Colour Library
p. 63 Eugene Fleury
Eugene Fleury
p. 64 Eugene Fleury
p. 66/67 Michael Holford Library
p. 68 Ronan Picture Library
p. 68/69 Ronan Picture Library
p. 70/71 Michael Holford Library
p. 72 Ronan Picture Library
p. 73 Ronan Picture Library
Ronan Picture Library/E. P. Goldschmidt & Co. Ltd.
p. 74 R. Swinfen/D. Nicholson
By permission of the Smithsonian Institution
p. 75 Michael Holford Library
p. 76 Geoffrey Watkinson
p. 78 Sonia Halliday Photographs
p. 79 From the original photograph in the Wellcome Institute by courtesy of the Trustees
p. 80 By courtesy of the Oriental Institute, University of Chicago
University Museum, Pennsylvania
p. 81 Ronan Picture Library/E. P. Goldschmidt & Co. Ltd.
p. 82 Ronan Picture Library/E. P. Goldschmidt & Co. Ltd.
p. 83 University Museum, Pennsylvania
p. 84/85 Barnaby's Picture Library/Gunter R. Reitz
p. 86 William MacQuitty
p. 87 By courtesy of the Oriental Institute, University of Chicago
p. 88 Barnaby's Picture Library
p. 89 The Mansell Collection
p. 90 Ronan Picture Library
p. 91 Michael Holford Library
p. 92 Anne Horton
p. 93 From the original chart in the Wellcome Institute by courtesy of the Trustees
p. 94/95 John Smith
p. 95 Michael Holford Library
p. 97 Popperfoto
p. 98 Spectrum Colour Library
p. 99 Spectrum Colour Library
p. 100 Sonia Halliday Photographs
p. 101 Popperfoto
p. 102 Staatliche Museen Berlin
p. 102/103 The Mansell Collection
p. 104/105 The Mansell Collection
p. 106 Geoffrey Watkinson
p. 107 Peter Weller
p. 108 Popperfoto
p. 110 Michael Holford Library
p. 112 Ronan Picture Library/E. P. Goldschmidt & Co. Ltd.
Ronan Picture Library/Royal Astronomical Society
p. 114 Ronan Picture Library

p. 115 John Smith
p. 116 Ronan Picture Library
Ronan Picture Library
Ronan Picture Library
p. 117 Ronan Picture Library
p. 118/119 John Smith
p. 120 Ronan Picture Library
p. 121 Ronan Picture Library